SpringerBriefs in Plant Science

SpringerBriefs present concise summaries of cutting-edge research and practical applications across a wide spectrum of fields. Featuring compact volumes of 50 to 125 pages, the series covers a range of content from professional to academic. Typical topics might include:

- A timely report of state-of-the art analytical techniques
- A bridge between new research results, as published in journal articles, and a contextual literature review
- A snapshot of a hot or emerging topic
- An in-depth case study or clinical example
- A presentation of core concepts that students must understand in order to make independent contributions

SpringerBriefs in Plant Sciences showcase emerging theory, original research, review material and practical application in plant genetics and genomics, agronomy, forestry, plant breeding and biotechnology, botany, and related fields, from a global author community. Briefs are characterized by fast, global electronic dissemination, standard publishing contracts, standardized manuscript preparation and formatting guidelines, and expedited production schedules.

More information about this series at http://www.springer.com/series/10080

Jillian De Gezelle

Q'eqchi' Maya Reproductive Ethnomedicine

Springer

Jillian De Gezelle
Dept. of Plant & Microbial Biology
North Carolina State University
Raleigh
North Carolina
USA

ISSN 2192-1229 ISSN 2192-1210 (electronic)
ISBN 978-3-319-10743-1 ISBN 978-3-319-10744-8 (eBook)
DOI 10.1007/978-3-319-10744-8
Springer Cham Heidelberg New York Dordrecht London

Library of Congress Control Number: 2014952588

Printed on acid-free paper

Springer is part of Springer Science+Business Media (www.springer.com)

Preface

The Q'eqchi' Maya of Belize have an extensive ethnopharmacopoeia of medicinal plants used traditionally for reproductive health and fertility. Ethnobotanical research was carried out in the Q'eqchi' communities of the Toledo District of Southern Belize from 2007–2011 on medicinal plant species used for reproductive health. Data was gathered primarily through semistructured interviews and plant collecting trips with 6 traditional healers, 3 midwives, and 12 female herbalists. The Belizean Q'eqchi' are utilizing more than 60 plant species for reproductive health treatments, with the most species from the black pepper family, Piperaceae.

Ten species were selected for investigation of their estrogenic activity using a reporter gene assay: *Clidemia crenulata* Gleason, *Drymonia serrulata* Jacq. (Mart.), *Gouania lupuloides* (L.) Urb., *Miconia oinochrophylla* Donn. Sm., *Mimosa pudica* L., *Piper jacquemontianum* Kunth, *Piper peltatum* L., *Psychotria acuminata* Benth., *Psychotria poeppigiana* Müll. Arg., and *Tococa guianensis* Aubl. These plants are used to treat female infertility, male infertility, menopausal symptoms, heavy menstruation, uterine fibroids, *k'uub'sa'* (a Q'eqchi' womb disorder), for miscarriage prevention, for use as female contraception, and for male contraception. Methanol extracts of the leaves of all species were assayed, as well as the stems of *G. lupuloides*, roots of *M. pudica*, and the roots of *P. peltatum*. All the extracts displayed estrogenic activity, except for *M. pudica* roots and *P. jacquemontianum* leaves, which were both cytotoxic to the MCF-7 breast cancer cell line. Nine of the species assayed were estrogenic, four of the species were also antiestrogenic, and two of the extracts were cytotoxic to the MCF-7 cell line.

Women's healing traditions are being lost in the Q'eqchi' communities of Belize at an accelerated rate, due to a combination of factors including: migration from Guatemala disrupting traditional familial lines of knowledge transmission; perceived disapproval by local biomedical authorities; women's limited mobility due to domestic obligations; and lack of confidence stemming from the devaluation of women's traditional knowledge. Medicinal plant knowledge is highly gendered with women and men commonly using different species in reproductive health treatments. Revitalizing women's healing practices is vital for maintaining the traditional knowledge needed to provide comprehensive healthcare for Belize's most remote indigenous communities.

Jillian De Gezelle

Acknowledgments

The blossoming and fruition of this project would not have been possible without the assistance, collaboration, and guidance of many individuals. I am very thankful for the support and advice offered by my mentors, colleagues, research participants, family and friends, without many of whom this work would not have been possible. Foremost, I would like to thank my graduate advisor, Dr. Michael Balick for his years of dedicated mentorship. I am incredibly grateful for the insightful guidance he has offered throughout my fieldwork in Belize and graduate studies, as well as his support of my professional development as an ethnobotanist.

I would like to thank the other members of my doctoral committee for their valued guidance and support as well: Dr. Ina Vandebroek of The New York Botanical Garden, Dr. Maryam Bamshad and Dr. Edward Kennelly of Lehman College, and Dr. Gail Mahady of the University of Illinois at Chicago. Thank you to the members of Dr. Mahady's laboratory for their collaboration in assaying plant extracts from this research. I would like to thank also Dr. Adam Kavalier for his phytochemical expertise, Dr. Nat Bletter and Dr. Lisa Offringa for their ethnobotanical expertise, Dr. Oren Tzfadia and Dr. Paul Matthews for their expertise in molecular biology, and naturopath Dr. Zoe Palmer Wright for her medical expertise.

I would like to thank some of the many individuals who have guided me in my research in Belize, foremost Don Victor Cal of the Belize Indigenous Training Institute. He as well as the traditional healers of the Q'eqchi' Healers Association have been my closest colleagues in Belize, and I am grateful to have had the opportunity to work closely with them. I have learned so much from their incredible collective wisdom, in particular from the revered Q'eqchi' Maya traditional healer, Don Francisco Caal, who is an expert in Q'eqchi' reproductive ethnomedicine.

Thank you to the other traditional healers of the Q'eqchi' Healers Association, Don Manuel Baqui, Don Emilio Cal, Don Lorenzo Choc, Don Manuel Choc, and Don Domingo Pop. I would also like to thank the following people for their contribution to the ethnomedical research, Ms. Dominga Ack, Ms. Francisca Ack, Ms. Adelina Ash, Ms. Bartola Avilez, Ms. Anita Caal Coc, Ms. Louisa Chiac Pacham, Ms. Juana Chub, Ms. Manuela Cus, Ms. Rosefina Cus, Don Juan Cus Coc, Ms. Jesus Ico, Ms. Diana Lara, Mr. Eduardo Lara, Ms. Eugenia Lara, Ms. Dolores Oh, Ms. Petrona Pan, Ms. Marcelina Te, and Ms. Laticia Zuniga.

Thank you to my research assistants and Q'eqchi' Mayan, Spanish and Garifuna interpreters, Maria Lara, Pedro Maquin, Lucy Alvarez, Fercia Maquin and Federico Caal. I would like to thank Tomas Caal for providing accurate Q'eqchi' Mayan spellings. My research assistant, Maria Lara has worked very closely with me over the years and become a dear friend. Her language skills, intellect and cultural understanding contributed greatly to this research. The entire Lara family, including Maria's parents and sisters, have helped with this work and invited me into their homes for many occasions. The lovely matriarch of the family Ms. Francisca Ack, Maria's grandmother, has graciously shared her knowledge and life experiences with me and made particularly meaningful contributions to this study.

I would like to thank Dr. Rosita Arvigo and Dr. Gregory Shropshire for kindly providing me with housing in Cayo and especially to Dr. Arvigo for guiding me during the early stages of my research on women's reproductive health and fertility. I am thankful to have had many of my first conversations on Belizean culture, politics and history with sociologist Melanie Santiago, and to be introduced to these topics through the perspective of such a strong, smart Belizean woman. Thank you to anthropologist Dr. Sophie Haines for her insight on Southern Belizean culture and Maya village life.

Thank you to Eric Stannard, Botany Editor at Springer, for his interest in ethnobotanical research and my study in Belize, as well as for providing guidance in the early stages of manuscript preparation. I am grateful also for all the assistance and guidance that Springer Developmental Editor Margaret Burns has provided throughout the process.

I am grateful for funding from The City University of New York including the Chancellor's Fellowship, the Writing Fellowship, and the Doctoral Research Grant Award; The Garden Club of America for the Anne S. Chatham Fellowship in Medicinal Botany; as well as the Institute of Economic Botany and the Graduate Program at The New York Botanical Garden.

Contents

1 Introduction 1
1.1 Ethnobotany and Ethnomedicine 1
1.2 Purpose and Scope of Research 1
1.3 Research Questions 2
1.4 Hypotheses 2
1.5 Background 2
1.5.1 Natural and Cultural History of Study Site 2
1.6 Traditional Medicine in Belize 8
1.6.1 Belizean "Bush Medicine" 8
1.6.2 Traditional Healing in the Belizean Q'eqchi' Community 9
1.6.3 Belize Indigenous Training Institute 12
1.6.4 Maya Spirituality 13
1.6.5 Culture-Bound Illnesses in Belize 15
1.6.6 Hot–Cold Duality in Maya Medicine 17
1.6.7 Maya Diagnostics 20
1.6.8 Traditional Q'eqchi' Healing Modalities 21
1.6.9 Medicinal Plant Harvest and Use 21
Conclusions 24
References 24

2 Traditional Q'eqchi' Maya Reproductive Ethnomedicine 27
2.1 Methodology 28
2.1.1 Study Site and Timeline 28
2.1.2 Research Ethics 28
2.1.3 Interviews 29
2.1.4 Plant Collection and Identification 30
2.2 Results and Discussion 30
2.2.1 Disorders of the Womb 31
2.2.2 Menstruation 34
2.2.3 Fertility, Infertility, and Fertility Regulation 40
2.2.4 Conception and Pregnancy 47

2.2.5 Birth and the Postpartum Period ... 51
2.2.6 Breastfeeding and Weaning ... 59
2.2.7 Menopause ... 61
2.2.8 Men's Reproductive Health ... 63
2.2.9 Infant Health ... 64
2.2.10 Culturally Specific Infant Conditions ... 66
Conclusions ... 68
References ... 69

3 Use of Phytoestrogens in Q'eqchi' Maya Ethnomedicine ... 71
3.1 Traditional Maya Medicine ... 71
3.2 Ethnomedically Directed Bioactivity Investigation ... 71
3.3 Phytoestrogens and Their Therapeutic Importance ... 72
3.4 Hormone Replacement Therapy (HRT) and the Women's Health Initiative ... 72
3.5 Q'eqchi' Maya Reproductive Ethnomedicine ... 73
3.6 Methodology ... 74
3.6.1 Belize Indigenous Training Institute ... 74
3.6.2 Ethnobotanical Research ... 74
3.6.3 Species Selection and Plant Collection ... 74
3.6.4 Memorandum of Agreement for Laboratory Research ... 75
3.6.5 Medicinal Plant Extraction ... 75
3.6.6 UIC/NIH Center for Botanical Dietary Supplements Research ... 75
3.6.7 Bioassay Materials ... 76
3.6.8 Cell Culture ... 76
3.6.9 Transient Transfections and Reporter Gene Assays ... 76
3.6.10 Cell Proliferation Assay ... 77
3.7 Results and Discussion ... 77
3.7.1 Cytotoxicity Results ... 77
3.7.2 Estrogenic and Antiestrogenic Activities ... 77
3.7.3 Fabaceae: Mimosoideae ... 77
3.7.4 Gesneriaceae ... 78
3.7.5 Melastomataceae ... 79
3.7.6 Piperaceae ... 81
3.7.7 Rhamnaceae ... 83
3.7.8 Rubiaceae ... 83
Conclusion ... 84
References ... 85

4 Q'eqchi' Women's Lives, Healthcare and Cultural Loss ... 87
4.1 Q'eqchi' Maya Women's Lives in Belize ... 87
4.2 Access and Choices in Reproductive Healthcare ... 90
4.3 Women's Healing Traditions ... 91
4.4 Methodology ... 92

4.4.1 Research Objectives 92
4.4.2 Interviews with Q'eqchi' Women 93
4.4.3 Interviews with Midwives, Herbalists and Healers 93
4.5 Results and Discussion 93
4.5.1 Q'eqchi' Women of a Toledo Village 93
4.5.2 Q'eqchi' Women's Health Perspectives 94
4.5.3 Q'eqchi' Choices and Perceptions of Pregnancy and Birth 95
4.5.4 Q'eqchi' Women's Traditions 97
4.5.5 Gender Differences in Healing and Plant Use 101
4.5.6 Q'eqchi' Midwives and Healers in Belize 102
4.5.7 Loss of Female Healers in the Belizean Q'eqchi Community 103
4.5.8 Q'eqchi' Migration 107
Conclusions 109
References 110

5 Conclusions 113
5.1 Q'eqchi' Reproductive Ethnomedicine 113
5.2 Estrogenic Plant Use in Q'eqchi' Reproductive Ethnomedicine 117
5.3 Q'eqchi' Women's Healing Traditions in Belize 118
References 121

Appendix 123
Appendix I. Interview Questions 123

Chapter 1
Introduction

1.1 Ethnobotany and Ethnomedicine

The relationship between plants and people has its origins in the beginning of our coexistence and has evolved into a remarkable diversity of manifestations. Plants now have an essential role in such diverse areas of our lives as agriculture, medicine, religion, construction, textiles, the visual arts, and music. The multidisciplinary study of the relationship between people and plants is known as ethnobotany. This includes medical ethnobotany, the study of the use of medicinal plants by people for healing. Medicinal plant use cannot be comprehensively studied or fully understood without the complementary investigation of the culture of the traditional medical system. The study of a culture's traditional medical practices is known as ethnomedicine, and includes beliefs on health and disease, diagnostics, health-related practices, and all aspects of medical treatments. My research unites the disciplines of medical ethnobotany and ethnomedicine to comprehensively study Q'eqchi' reproductive ethnomedicine with a focus on the medicinal plants utilized in this branch of traditional Maya medicine.

1.2 Purpose and Scope of Research

The purpose of my interdisciplinary research in Belize was threefold: first, to research and document practices and beliefs related to Q'eqchi' Maya reproductive ethnomedicine, with particular focus on traditional plant-based remedies related to reproductive health and fertility regulation; second, to assay extracts of a selection of the medicinal plants used in Q'eqchi' reproductive ethnomedicine for relevant bioactivity, as a means to further validate traditional Maya medicine and Maya traditional wisdom; and third, to investigate the causes and consequences of the significant decline in female practitioners of Q'eqchi' Maya traditional healing and midwifery in the indigenous communities of Belize.

J. De Gezelle, *Q'eqchi' Maya Reproductive Ethnomedicine*,
SpringerBriefs in Plant Science, DOI 10.1007/978-3-319-10744-8_1

1.3 Research Questions

1. What ethnomedical treatments, including medicinal plant remedies are used for reproductive health in Q'eqchi' traditional medicine?
2. What bioactivity may account for the medicinal efficacy of plants used traditionally by the Q'eqchi' people of Belize for the treatment of reproductive health-related conditions?
3. Why is there a decline in traditional female healers and midwives in the Q'eqchi' communities of southern Belize?

1.4 Hypotheses

1. Q'eqchi' traditional medicine has an extensive ethnopharmacopoeia of medicinal plant species used for reproductive health, as well as culturally specific ethnomedical practices.
2. Traditional Q'eqchi' medicine treats conditions related to hormone imbalances with plants containing compounds whose bioactivity can be demonstrated in standard hormone bioassays.
3. Women's healing traditions are being lost in the Q'eqchi' communities of southern Belize due to a combination of factors related to their emigration from Guatemala and the sociocultural setting of Belize.

1.5 Background

1.5.1 Natural and Cultural History of Study Site

Belize

Belize is a small Caribbean nation located in Central America, bordered by Mexico to the north and Guatemala to the south and west. The eastern edge of Belize meets the Caribbean Sea. Belize has the second-longest barrier reef in the world, and the longest in the northern hemisphere, extending 220 km, with 79 % of the Mesoamerican Reef System in its territorial waters (Hartshorn et al. 1984; Jacobs and Castañeda 1998). Belize is comprised of 22,963 km^2 of land area that includes 689 km^2 on 450 cayes off of its shores (Hartshorn et al. 1984). Precipitation in Belize varies from a 4-month dry season with less than 1300 mm of rain per year in the north, to a shorter dry season in the south with over 4000 mm of rain per year creating a range of ecological niches, in a gradient from north to south getting increasingly wet. This small country is home to 576 species of birds, 288 species of Lepidoptera, 163 species of mammals, and 122 species of reptiles (Jacobs and Castañeda 1998). Seventy percent of the country is estimated to still be under forest cover today, though up until the early 1980s 93 % of the country was classified as "forest land" (Balick and O'Brien 2004).

An anomaly among the Spanish-speaking countries in Central America, Belize was once a British colony (then known as British Honduras), and English is the official language. An English Creole—Kriol—is the language spoken ubiquitously and the culture has a strong Caribbean influence and association. It is at once Central American, Caribbean, and most significantly,uniquely Belizean. Belize is ethnically very diverse, with a population comprised of Kriol (Creole), Garinagu (Garifuna people), Q'eqchi' Maya (Kekchi), Mopan Maya, Yucatec Maya, East Indians, Mestizos, Mennonites, Taiwanese, Chinese, Lebanese, American and European expatriates, and others. The country has the lowest population density of all the Central American countries, and few major roads. Thus, the country has maintained much of its natural habitat and the biological diversity it holds.

The Belize Ethnobotany Project began in 1988, which was the largest ethnobotanical and floristic inventory ever undertaken of this country. Central to the project was an extensive survey of ethnomedical practices and beliefs, and medicinal plant knowledge. This project also produced and published the *Checklist of the Vascular Plants of Belize* (Balick et al. 2000). The scientific team documented 3408 species of vascular plants, 1219 genera, and 209 families known to the country. *Schippia*, a monotypic genus along with 41 other species are endemic to Belize, comprising 1.2 % of the native and naturalized flora of the country. Of the total flora, including cultivated taxa, 38 % were found to be considered useful. In a country of such cultural and biological diversity, knowledge of useful and medicinal plants is a common thread running across cultures and is a part of the collective Belizean national heritage.

Toledo District

Belize has six districts, with the Toledo District being the southernmost. Punta Gorda (locally called "P.G.") is the district town for Toledo, and has a population of around 5000 people. P.G. has bus service out to the 30 or so villages in the Toledo District, which has a population total of ca 30,000 (Statistical Institute of Belize 2011). Toledo is the most remote district; Punta Gorda is a coastal town (Fig. 1.1) located at the end of the Southern Highway and is approximately 7 h by local bus from Belize City.

Toledo has the largest indigenous population in the country, and the majority of Q'eqchi' Maya, Mopan Maya, and Garifuna people living in Belize reside in this southernmost district. Toledo has the lowest population density, the largest average household size, and the least access to electricity and public water in the country, as well as many other indicators of disadvantage among the Belizean population. Furthermore, Toledo has the greatest disparity between males and females with regard to education levels and employment rates (Statistical Institute of Belize 2007). Q'eqchi' women I spoke with during my research often said their biggest struggle

Fig. 1.1 Aerial view of Punta Gorda Town, Toledo District, Belize

was finding employment with limited English and limited employable skills, and that their greatest hope for their daughters is that they are able to obtain a better education and have more career opportunities than they have had.

The Toledo District receives some of the highest rainfall in Belize, with mean annual rainfalls of 2500–4000 mm. During the wet season, from June to December, there is rain nearly every day, though it is much more variable during the dry season from January to May. Ecological Life Zones of Toledo are classified as Subtropical wet to Tropical wet (Hartshorn et al. 1984). Toledo is divided into two regions naturally—upland hills and lowland coastal area. Four major rivers run through Toledo, from north to south—the Rio Grande, Moho, Temash, and Sarstoon rivers—the latter three originating in Guatemala. The extensive mangrove swamp limits the settlement of coastal areas to a few higher points like Punta Gorda, Punta Negra and Barranco, which have had predominantly Garifuna populations (Wilk 1997; Woods et al. 1997), though today Punta Gorda especially has a very diverse population of Kriol, East Indian, Maya, Asian, and expatriate people. Q'eqchi' settlements have been located farther inland at the edges of hills where there is access to better-drained upland soils or along river banks (Wilk 1997; Woods et al. 1997).

Forests of the Toledo District are characterized by tree species such as banak (*Virola koschnyi* Warb.), quamwood (*Schizolobium parahybum* (Vell.) S.F. Blake), rosewood (*Dalbergia stevensonii* Standl.), santa maría (*Calophyllum brasiliense* Cambess.), and yemeri (*Vochysia hondurensis* Sprague) (Hartshorn et al. 1984).

Fig. 1.2 Yard in Punta Gorda Town with various plants and homemade ice for sale

Other conspicuous species include the cohune palm (*Attalea cohune* Mart.), ceiba (*Ceiba pentandra* (L.) Gaertn.), gumbolimbo (*Bursera simaruba* (L.) Sarg.), and allspice (*Pimenta dioica* (L.) Merr.).

Historically, exports from Toledo have been primarily timber and agricultural products, including mahogany, logwood, chicle, rubber, sugar, rum, bananas, coffee, and cacao. However, as Wilk (1997) put it "Toledo still looks like a frontier because each time it has been penetrated the land and its people have been so thoroughly and efficiently exploited that capitalist ventures left nothing behind, except the workers themselves." He goes on to explain that without government investment in infrastructure, such as roads and communications, only industries that extract resources can profit, which perpetuates the cycle of underdevelopment. Wilk points out that, "in the 1860s Toledo was no more backward or marginal than any other part of rural British Honduras. Today, through this cycle, it has become underdeveloped." Toledo is frequently referred to as "the Forgotten District" in discussions relating to its historical underdevelopment, relative to the rest of the country (Crooks 1997). In order to make a living, many Belizeans living in Punta Gorda and Toledo need to be innovative and diversify their means of income (Fig. 1.2).

Q'eqchi' Maya

The Q'eqchi' are one of three Maya ethnic and language groups residing in Belize, along with the Mopan Maya and the Yucatec Maya. Originally the term "Q'eqchi'" referred to the Mayan language, not to an ethnic group, and any person who spoke this language fluently could be called an "*aj Q'eqchi'*" including Maya individuals originally from other places (Wilson 1993). The Q'eqchi' in Belize are a small portion of the whole Q'eqchi' population, which is estimated to be between 700,000 and 900,000 and largely reside in the Alta Verapaz Department of Guatemala. Nearly half of the population of Belize's Toledo District is Q'eqchi'—around 14,000 out of 30,000 (Statistical Institute of Belize 2011). The Q'eqchi' reside primarily in the villages of the Toledo District, as well as the Indianville neighborhood at the inland edges of the coastal district town of Punta Gorda. Many of the Q'eqchi' villages in Belize now have access to water and electricity, but some still do not. While some Q'eqchi' live in concrete or wooden houses with metal roofing, most Q'eqchi' live in traditional rectangular houses with hard-packed dirt floors, wooden boards for walls, and thatched roofs made from *cohune palm* (*Attalea cohune* Mart.) leaves or *bay leaf palm* (*Sabal* sp.) leaves.

Notable scholarly research on the Belizean Q'eqchi' Maya includes Richard Wilk's *Household Ecology: Economic Change and Domestic Life among the Kekchi Maya in Belize* (1997). The majority of Q'eqchi' people are small-scale, nonmechanized farmers who focus production on *corn* (*Zea mays* L.) and *black beans* (*Phaseolus vulgaris* L.) on cleared and cultivated land known as *milpa*, as well as *rice* (*Oryza sativa* L.) and *cacao* (*Theobroma cacao* L.) (Wilk 1997; Collins and Liukkonen 2002). Q'eqchi' men will often take wage labor jobs to supplement their income, and Q'eqchi' women (and some men) produce crafts for sale primarily to tourists. There are Q'eqchi' people, especially many younger individuals, who work for the local NGOs, are teachers, involved in politics, own businesses, and are scholars, doctors, and lawyers regionally and internationally.

Corn is the staple food of the Q'eqchi' and its production is treated with a spiritual reverence not given to any other crops in their culture. Rituals including all-night vigils of prayer, song, feasting, and offerings of *copal* (*Protium copal* (Schltdl. & Cham.) Engl.) incense, known as *pom* in Q'eqchi', are enacted to ensure harmony between the farmers and the *Tzuultaq'a* who are the spirits or deities of the forests and the hills (Wilk 1997). *Corn,* in the form of tortillas (*cua* in Q'eqchi'), tamales, or steamed dumplings (*poch* in Q'eqchi'), is typically combined with a meat or a vegetable in Q'eqchi' meals. Chicken, pork,or eggs are relatively common and are included as available and affordable. *Black beans* (*Phaseolus vulgaris* L.), *red kidney beans* (*Phaseolus vulgaris* L.), *pumpkins* (*Cucurbita* spp.), *okra* (*Abelmoschus esculentus* (L.) Moench), *calaloo* (*Amaranthus dubius* Mart. ex Thell.), *tomatoes* (*Solanum lycopersicum* L.), *avocados* (*Persea americana* Mill.), *plantains* (*Musa × paradisiaca* L.), *papaya* (*Carica papaya* L.), *pineapples* (*Ananas comosus* (L.) Merr.), and *citrus* (*Citrus* spp.) are common foods grown, exchanged, and eaten in Q'eqchi' villages. Spices used in Q'eqchi' cooking include *chiles* (*Capsicum* spp.), *garlic* (*Allium sativum* L.), *onion* (*Allium cepa* L.), *cilantro* (*Coriandrum sativum* L.), *culantro* (*Eryngium foetidum* L.), *achiote* (*Bixa orellana* L.), *ginger* (*Zingiber officinale* Roscoe), *allspice* (*Pimenta dioica* (L.) Merr.), and *wild basil* (*Ocimum campechianum* Mill.). *Cacao* (*Theobroma cacao* L.), *coffee* (*Coffea arabica* L.), and *corn* porridge are common drinks in Q'eqchi' households. Wilk (1997) explains, "the identity of corn as the true food, without which eating becomes snacking, does not keep the Kekchi from trying new foods and growing new crops. But

Fig. 1.3 Market day on Punta Gorda's Front Street where women and men from the villages sell produce, as the colorful village buses that brought them to town pass by

new foods and new crops are eaten and grown without ceremony, without sanctity. They may sustain the body, but not the heart."

The Toledo District is still characterized by rural agricultural production of fruits and vegetables brought from the villages to Punta Gorda Town, to as far as Belize City and the cayes. Many of the Toledo villages only have buses that travel to and from Punta Gorda Town on market days, which are Monday, Wednesday, Friday, and Saturday. The village buses are old school buses privately owned by individual families and they will leave from the village early enough to reach Punta Gorda around 6 a.m., and will return to the village around noon. Larger villages may have bus service daily and at additional times, such as a bus returning to the village later in the afternoon. Villagers travel to town primarily on market days, with some regularity, depending on their needs, and do their shopping, banking, and anything else they need to handle in town. Villagers who grow food to sell or purchase food to sell bring their produce on the bus to market and set up along Front Street in Punta Gorda Town (Fig. 1.3).

Wild harvested foods or "bush foods" are an important supplement to the Q'eqchi' diet, and add occasional nutritional and taste variety to the general diet, as well as serve as emergency foods during rare times of crop failure. Q'eqchi' wild harvested foods include the soft heart of the *cohune palm* (*Attalea cohune* Mart.)

known as *mokoch* or "cohune cabbage", cohune nuts, and the leaf shoots from *jippi jappa* (*Carludovica palmata* Ruiz & Pav.) (Wilk 1997). *Halau* (also known as *gibnut* in Kriol) is a popular hunted "bush meat" in Q'eqchi' villages, and peccary, brocket deer, armadillo, iguana, birds, fresh water fish, and snails (known as *jutes*) among other wild animals are eaten as well.

1.6 Traditional Medicine in Belize

1.6.1 Belizean "Bush Medicine"

In their research, Arvigo and Balick (1993) found that nearly every Belizean adult alive at that time had been raised on some form of home remedies by their parents or grandparents. Many Belizeans live in rural and forested areas and rely on Belize's biological diversity to contribute to many needs including medicine, food, and construction materials. Home remedy knowledge is widespread among Belizeans, even those residing in the few urban areas. Staiano (1981) explains that in Belize, most people treat minor illnesses themselves with "bush medicine," plants they or friends or relatives already have growing around their home. The knowledge of which plants are useful for which purposes is readily shared throughout the community. Therefore, in Belize through "the process of interaction, several ethnic groups seem to have adopted some of the drugs and therapies of other groups so that there now exists a rather large body of shared knowledge which includes information on both bush and patent medicines and might be referred to as 'home remedies'."

There are traditional healers and herbalists in the Q'eqchi', Mopan, Yucatec, Garifuna, Kriol, East Indian, Mestizo, and Mennonite communities of Belize. Most are elders, however, and few young Belizeans are taking traditional healing up as a professional pursuit. Balick and O'Brien (2004) explain that Belizean healers' knowledge has been developed through a long and close ancestral relationship with nature, but that outside cultural influences and land clearing have threatened the biocultural diversity upon which this knowledge is based. They reported that "herb gatherers now claim that they have to travel longer and longer distances to obtain medicinal plants as the diversity and availability of medicinal plants are being destroyed. It is evident also that knowledge about the use of plants is itself in greater danger of extinction than the plants themselves."

Traditional healing is referred to as "bush medicine" in Belize and traditional healers are known as "bush doctors," since they use medicinal plants from the "bush" or forest. Traditional healers in Belize have historically had specialized roles and functions. These included the Doctor–Priest/Priestess who has the power to deal with both physical and spiritual ailments, the Village Healer who served as the primary health care provider for their extended families, the Grannie Healer who raised their own children on home remedies and consulted with the Village Healer for ailments they could not successfully treat. Even more specialized were the Midwives, Bonesetters, Massage Therapists, and Snake Doctors (Arvigo and Balick

Fig. 1.4 Herb stall in Belize City selling mostly dried packaged herbs

1993). These highly specialized branches of traditional medicine are the least common in Belize today. In Belize City and a few of the larger towns in Belize, a market stall selling medicinal herbs can be found, often selling packaged herbs (Fig. 1.4).

Traditional healing in Belize and Central America is primarily a syncretic practice involving the mixture of ancient indigenous traditions and Christianity. Indigenous peoples were able to maintain their healing and spiritual traditions by combining those with the Christian practices brought by the conquest of Latin America. And certainly today in a diverse country with a shared language, traditional indigenous healing practices are influenced by other cultures' traditional knowledge as well as by biomedicine. Traditional knowledge has been transmitted orally until recent years when ethnobotanical research efforts began to record this information in Belize (Arvigo and Balick 1993).

1.6.2 Traditional Healing in the Belizean Q'eqchi' Community

Just a couple of generations ago the wealth of knowledge of medicinal plants in the Q'eqchi' community was much more widespread throughout members of the community. Today the deep knowledge is mostly held by a small number of male traditional healers, and even fewer women, mostly older, who often go unrecognized for their depth of knowledge, at the community level. Q'eqchi' people in Belize today are for the most part not using the services of traditional healers to meet all of their medical needs. Some Q'eqchi' will go to the traditional healer first for treatment, and if the treatment is not successful, they will then try the hospital. Others will try the hospital's treatment first, and if this is not successful, they will go visit a traditional healer for treatment. Other Q'eqchi' say they will try one or the other first depending on what the ailment is.

In the Q'eqchi' community a traditional healer who treats both physical and spiritual ailments is referred to as an *ilonel*, which literally translates to "watcher" (Collins and Liukkonen 2002) and they practice *ilok* (healing). Both males and females can be *ilonels* and a female healer may be referred to as an *ilonel ixq* (woman healer). Additionally there are spiritual guides (*guías espirituales* in Spanish) who handle all spiritual workings including cleansings and blessings, and may not necessarily practice much in the way of herbal medicine. Though the Spanish term is often used, spiritual guides may be referred to as *aj kamolbe re Maya jak* ("leader of Maya incense burning and prayer") or *jolomil aj Maya janel* ("leader/teacher of Maya ceremonies"). One male healer called himself a *remer*, who is someone who heals only through prayer, and this was perhaps indicative of the fact that he no longer hikes into the forest to collect medicinal plants due to his age and declining health. The two branches of traditional healing—physical and spiritual—are aligned and overlap, however *ilonels* are the experts in medicinal plants and the *guías espirituales* are experts in traditional Q'eqchi' spiritual matters.

Midwives are known as *ilol ixq* and traditionally specialize in reproductive health especially of women, pregnancy, birth, and newborn health. Three of the women who were interviewed for this study were midwives and one of them was no longer practicing. She said she would call herself a "woman healer" now. None of the other women from this study said they call themselves any kind of healer, but I refer to them as herbalists. Some specifically stated, "I am not a healer, I treat only in the family." One of these same women, however, treats high blood pressure, diabetes, worms, diarrhea, vomiting, coughing, headache, delayed menstruation, when babies have "wind," *susto* (*kaanil* in Q'eqchi', "fright" in English), rashes, sores, and more. Although there are very few Q'eqchi' midwives living and practicing in Belize, the practice has been maintained in the Q'eqchi' communities of Guatemala (Michel et al. 2007). Chapter 4 discusses this loss of Q'eqchi' women's healing traditions in Belize in more depth.

In my study, the average age of the male healers interviewed was 68 years old and ranged from 55 to 85 ($N=6$). The average age of female midwives and herbalists was 50 years old and ranged from 22 to 75 ($N=15$). Note that many people do not know their exact age, especially the elder generation, but even some women in their reproductive years have forgotten their age and maybe have never known for sure. All of the men and half of the women were born in Guatemala, the other half of the women were born in Belize.

Most of the men and some of the women have not been formally educated at all because none was available, or they otherwise did not have the opportunity, perhaps because of the cost. One male traditional healer attended literacy classes while living in Guatemala. Some of the female herbalists attended school in Belize finishing between Standard 4 (equivalent to 6th grade in the US system) and 4th form (equivalent to 12th grade in the US system). One of the midwives came out of school as a young girl but later was trained in midwifery for 3 years by Mexican midwives at a clinic in Guatemala. In general, the younger individuals had more formal education, and the elder healers, midwives, and herbalists had less formal education. Schools were not available in all villages in the Petén region of Guatemala or the Toledo District of Belize when many of the older individuals were of school age. Many of the older individuals speak only Q'eqchi', and perhaps a little Spanish. Some of the younger individuals also spoke Kriol/English, Mopan and/or Spanish in addition. One woman spoke all four languages.

The healers interviewed typically had one or two family members who had practiced, still practice, or are training to practice traditional medicine. These included a father, a stepfather, an uncle, one or two brothers, and a brother-in-law. Two healers had been training their own sons, who were not yet working as healers. Women herbalists and healers who were interviewed also typically had a family member who had been a healer or herbalist, with whom they typically trained. One midwife learned much of what she knows about midwifery from her deceased husband who was a traditional healer. Herbalists learned from sources including grandparents and parents, a mother who also only treated within the family, neighbors and friends, learned from an uncle or husband who was a traditional healer, from elders in the community, and traditional healers who had treated them. Two herbalists had sisters who had also been midwives. It seems the interest in traditional medicine, herbalism and midwifery runs in particular families. Furthermore, serious apprenticeships with traditional healers who are not family members can be relatively expensive, which may be why people who have an opportunity to learn from a family member for free are the most likely to learn about traditional healing.

Traditional healers when questioned would typically answer that they "treat everything" or "heal almost any illnesses or cases" or "physical as well as spiritual ailments." Women who are experienced herbalists typically answered with a string of several ailments that they have treated, which cover a wide range of conditions, yet they did not consider themselves healers. Two traditional healers who also specialize in spiritual ailments discussed their practice. One said, "I only help and cure people. No *obeah* (sorcery)." He does do "love spell" type work though, to help someone get a certain person's love, or to help someone regain their spouse's fidelity, for example. The other healer said that aside from physical ailments like when people cannot go to the bathroom or when they have excessive pain in their bones and joints, he can treat when people fall or are frightened and need their spirit drawn back into their body, and when a woman has trouble giving birth due to witchcraft that causes the child to be bound to the woman.

One woman who used to be a midwife stopped assisting during deliveries several years ago because of her worsening eyesight, but did attend to births for 12 years. Now she does massage and positioning the baby in the womb. She sends the mothers to the hospital for their births, and once they are home from the hospital she massages the mother and baby. She can also turn breech babies in the womb, using external cephalic version, a practice of manual manipulation of the positioning of the fetus, common to the skill sets of many midwives in developing countries (Jordan 1984). She says she treats heavy menstruation, back pain due to heavy menstruation, fallen fontanel in babies, and the complaints of older age women. She also treats some general ailments like stomach ulcers. Another midwife says she is the only midwife in her village. She does massage during pregnancies, she delivers babies, she treats menstrual problems, "mole drop" (fallen fontanel) in babies, babies' and kids' ailments, and teenagers with menstrual problems.

These two midwives had both been practicing for 20 years. The male traditional healers had been practicing for an average of 35 years, with a range of 12–52 years of practice ($N=6$). All the traditional healers treat women, men, and children in their practices, and said that their patients must all hear from previous patients who have been healed by them. People recommend their treatment and the word spreads. The herbalists for the most part only treat within their close circle of family, friends, and neighbors.

This is a part-time occupation for all the traditional healers, seeing a few patients a week, on average. Most of the male healers also do farming on their *milpas* to earn their living. Some are also employed doing building, cleaning yards, and doing other labor. Some of the male healers who are also a part of the Q'eqchi' Healers Association (QHA) are employed part-time at the Itzama Medicinal Garden, tending to the medicinal plants, chopping the yard and doing other maintenance of the site. Occasionally they are paid to give a tour there or are paid to collaborate for interviews and plant collections with researchers, such as myself. According to one healer, his patients just pay BZ$ 5 or 10 (US$ 2.50 or 5) or not at all, but he does not ask for more, only whatever they can pay. Most others operate this way as well—their payment is whatever the patient or family can pay, sometimes nothing. One older single woman who is an herbalist also has a piece of land that she grows produce on to sell at the market in Punta Gorda, in Indianville, and in her own village. To make money some of the other women would raise livestock, or find occasional work helping other people with their washing, cooking, and planting.

1.6.3 Belize Indigenous Training Institute

In Belize I collaborated with the Belize Indigenous Training Institute (BITI), a Belizean nonprofit organization, incorporated in 1998 (Pesek et al. 2007). Under BITI operates the only formally organized group of traditional healers in Toledo, the QHA. Currently, six male Q'eqchi' healers work together in the QHA to maintain a medicinal plant forest garden, share traditional medical knowledge with each other, and revitalize traditional Maya practices in their communities. These male healers confirmed my observation that women's traditional medical knowledge is particularly endangered in the community and that female traditional healers, who are recognized as such, are virtually nonexistent in the Q'eqchi' communities of Belize. The Belize Indigenous Training Institute shared my goals of: (1) dispelling misconceptions about traditional healers and Maya medicine and returning prestige to the practice; (2) further validating Q'eqchi' traditional wisdom through verification of medicinal qualities of their plant medicines with relevant bioassays and literature research; (3) collaborating with the local hospital and clinics to provide culturally competent and comprehensive healthcare to the people of Toledo; (4) and revitalizing the tradition of Q'eqchi' Maya women's medicine in Belize.

The Belize Indigenous Training Institute maintains a 50-acre medicinal forest garden known as Itzama, a name that means the home of the Maya god of wisdom (Itzamna) and a place of spiritual and herbal healing (Pesek et al. 2007). The land has a small cleared area with a couple of thatched open-air buildings for meetings and events and an area of sunny gardens. The rest of the land is forested with part running along the Golden Stream River. The healers cultivate medicinal plants under the canopy, transplanted from other areas of Toledo that were destined to be burned or cut down. All of the healers agree that they have to travel farther now than ever before to find the plants used in their healing practices, so the living plant collections at Itzama provide a nearby source of medicinal plant material. The healers are tending to over 100 species of medicinal plants used in Q'eqchi' traditional medicine, grown in their respective habitat niches (Pesek et al. 2007). The plants in the garden are used to promote healing and traditional practices through education and provide an additional source of plants for the healers, and is intended eventually to be a source for the production of phytomedicines or other products (Bourbonnais-Spear et al. 2006).

1.6.4 Maya Spirituality

Traditional Maya spirituality is integral to traditional Maya healing. Prayer and ritual accompany healing treatments and medicinal plant collections. Healers feel that the plant medicines will not be effective in healing the patient without the accompanying spiritual work. *Guías espirituales* practice spiritual cleansings, spiritual healings, and Maya traditional religious ceremonies specifically. Traditional healers, or *ilonels*, also have spiritual components to their healing practices, in treating spiritual as well as physical ailments. *Don* is the Spanish prefix used to denote respect for a male elder, and is utilized when referring to *ilonels*. Don Francisco, a revered *ilonel*, explained that his "Maya day" (the day he was born on in the Maya calendar) is 8th winds, which is why he is an adept healer. *Ik* is the Q'eqchi' word for wind, which represents cleaning, clearing, and cooling, all of which he associates with healing.

Today's Maya spirituality is a syncretic blend of ancient Maya religion and Christianity (see Fig. 1.5). Regarding the Quiché Maya of Guatemala, Cosminksy (1994) explains that most people are Catholic, and that they are either practicing a reformed Catholicism associated with the Catholic Action movement, or a "syncretic form based on sixteenth century Catholicism and Mayan Indian influences. This latter includes the use of the Mayan ritual calendar and shamans, worship of *El Mundo*, the essence or spirit of the earth, and *Aire*, the air, combined with the worship of God, Jesus, Mary, and the Saints."

During many Q'eqchi' traditional ceremonies, four colors of candles are used: red, which is associated with the east and sunrise; black, which is associated with the west and sunset; yellow, which is associated with the south and the strength of

Fig. 1.5 Maya woman's Easter altar with package of copal resin incense next to the Christian cross and statue, with offerings of traditional caldo, beans, corn tortillas, and corn *poch*

the waters; and white, which is associated with north and the strength of the air. Red, black, yellow, and white represent the four corners, and sometimes green and blue candles are also used which represent the leaves and the sky, respectively. The four corners are blessed in the ceremony, blessing spirits in each of the four corners, alive and dead and associated with trees and other components of the environment. *Copal* (*Protium copal* (Schltdl. & Cham.) Engl.) resin incense, called *pom* in Q'eqchi', is burned during ceremonies and is believed to unite the four corners and the soil. *Copal* is burned at nearly all Q'eqchi' rituals; the smoke of which carries prayers to the *Tzuultaq'a*, spirits of the forested hills and mountains (Wilk 1997; Wilson 1993), and to the four directions. Sugar is poured onto the ground in the sign of a circle, divided by an equal armed cross to create four equal quadrants. The spiritual guide (Fig. 1.6) will lay the candles and incense out on top of this along with *ensarte* (unidentified), which is tree resin and *cascarita* (unidentified) bark, which are also burned to call the spirits.

The *Tzuultaq'a*, spirits of the mountains, can be either male or female, have a human form and live in a 'house' or cave deep inside the mountain they inhabit. Only one *Tzuultaq'a* resides in each mountain, considered the "owner" of that mountain. Collectively the *Tzuultaq'a* own everything on the surface of the land, and are the guardians of people, plants, and animals of the forest, and the original owners of corn. Q'eqchi' elders know their *Tzuultaq'a* through their dreams. People have a reciprocal relationship with the *Tzuultaq'a* in which people must feed the mountains, not just take of its bounty and plant on its land without recompense. So every spring, Maya who maintain the ancestral spiritual agricultural practices will travel to the

Fig. 1.6 Q'eqchi' spiritual guide, Don Emilio, giving a spiritual blessing at the ruins of Tikal, in Guatemala

caves in the mountains to petition the *Tzuultaq'a* (Wilson 1993). Wilson (1993) explains how:

> In elders' dreams, the *Tzuultaq'as* demand more food in the form of copious amounts of sticky resinous incense, candles, fermented maize drink, turkey blood and raw cacao beans…Elders dream about the *Tzuultaq'as* frequently before the maize planting, and these dreams act as a justification for sacrifices in the caves.

1.6.5 Culture-Bound Illnesses in Belize

Culture-bound syndromes or illnesses are those that are specific to a culture, locality, or region and therefore may not be well understood or even accepted in other medical systems. In Belize, widely known culture-bound illnesses include *susto* ("fright"), *envidia* ("envy"), *tristeza* ("sadness"), *pesar* ("grief"), *mal de ojo* ("evil eye"), and *mal vientos* ("evil winds"), many of which are found throughout Latin American cultures (Arvigo and Balick 1993). A study on beliefs and practices related to *susto* in Guatemala, Mexico, and Texas, showed that a single description of the illness exists across these Hispanic cultures with strong similarities in causes and symptoms, however variation in treatments exist (Weller et al. 2002). In *The DSM-IV-TR: Diagnostic and Statistical Manual of Mental Disorders*, a culture-bound syndrome in the realm of mental illnesses refers to "recurrent, locality-specific patterns of aberrant behavior and troubling experience that may or may not be linked to a specific DSM-IV diagnostic category" (American Psychiatric Association 2002). The culture-bound syndromes appear in an appendix in the DSM-IV, which is undoubtedly an incomplete list. There has been recent debate with one side insisting the appendix should not be included at all in the next edition, the forthcoming DSM-V. The other side of the debate feels that all mental illnesses are culture bound and that some illnesses that Westerners consider universal are not found in other cultures at all.

Traditional Q'eqchi' healers first determine if their patient's illness is of spiritual or physical nature before deciding how to proceed with the appropriate treatment. Most do so through feeling the pulse. Traditional healer Don Francisco can tell if a person has a spiritual ailment through the pulse of his patient because it feels like the heartbeat of someone scared, pumping fast. Spiritual ailments can be caused by coming in contact with a spirit, particularly if it is an evil one, but this spirit may just be that of an ancestor. Spirits are believed to travel in the winds and embody the winds. Contact with spirits occurs in remote parts of the forest and in abandoned houses, often at night, but can happen anywhere at any time.

Another traditional healer explained that an evil spirit is considered the same as evil winds. He knows when someone has been afflicted by this because their eyes look upwards and they are sweaty and pale. He observed that in this case their pulse beats slowly. *Vientos* (winds) refers to the same illness in Q'eqchi' as "epilepsy" (*etajtyajel* in Q'eqchi') he explained, because this is caused by evil in the wind. "Fits" or *ataques* are caused by spirits, for instance if a person walks into an abandoned house and the spirit wants to play with their body, if someone goes to the forest alone, or if women or children are left home alone at night. A plant called *papim* (unidentified) because the leaf resembles a butterfly is used to treat these "fits." He says the remedy is made and placed under the dew because the dew that drops in is part of the remedy; it helps cool the head. The patient is to drink this at 5 a.m.

Nine white *rose* (*Rosa* sp.) flowers are mashed in a bath with nine local *limes* (*Citrus aurantifolia* (Christm.) Swingle) and nine cloves of *garlic* (*Allium sativum* L.), and *ruda* (*Ruta* sp.), and the water is used to bathe a person to protect them from spirits and evil eye. One woman said many people in her village carry out numerous spiritual practices to protect the children from spirits and very small men that are believed to take the children. *Xeelik* are spirits that people see sometimes often in the form of little men, that can cause sickness. They exist in the forest and by riversides. This sickness would be treated with plants and prayers. There are different symptoms for when a ghost sees someone and this would be treated by burning *garlic* so that the person can inhale, according to one healer.

When a child is sick with *susto* or "fright", they cut the hair of the others in the household and burn the hair, holding the child over the smoke. This is done with hair from all the other members of the household when it is unclear who frightened the child. This is something that traditional Q'eqchi' women do, one of the herbalists explained. During the interview with a traditional healer, a young boy, apparently the healer's grandson, who was lying in the hammock nearby started kicking his leg, and jostling the hammock. The healer went over to the boy and gingerly lifted his arm to feel his pulse. "His illness is coming back," he said. He covered the boy's head with a clear liquid out of an emptied glass Nescafe jar. He told my interpreter and me that the boy is haunted by the spirit of his aunt who died 2 years ago. The boy was calmed by the healer bathing his head. Another healer says his child was attacked by an evil spirit. He used *romero* (rosemary: *Rosmarinus officinalis* L.), *kaxla pum* (unidentified), *assune* (unidentified), and "Florida water." He bathed his son with this and had him drink some.

Don Francisco explained that sometimes if a woman "catches spirit," she might not be able to get pregnant. However, some women have strong immune systems and can be healed from this quickly and regain their fertility. According to another healer, some women experience secondary amenorrhea (menstruation stops, before menopause) because "menstruation becomes hard because of *susto* ("fright") and the women need medicinal plant teas to soften the hard blood." Some Q'eqchi' feel that for the first three nights postpartum the lights should be left on to protect from evil and keep spirits away from the new baby and mother.

Spiritual ailments can also be caused by someone practicing witchcraft with mal intent, often referred to as *obeah* in Belize. There are a lot of problems with witchcraft in the villages, according to one herbalist. There are what are known as *obeah* pregnancies, in which a woman is cursed and an evil spirit is called into her body by someone who wishes to cause her harm. Her body goes through what seems like a pregnancy, but she would not have been biologically impregnated. When a woman has an *obeah* pregnancy, one traditional healer says that he can feel her blood is like a chain, like the blood that is moving feels like beads of a chain moving. Similarly, in northwest Argentina (Hilgert and Gil 2007) there is the belief that if plants are used successfully to induce an abortion, then the pregnancy may have been caused by a supernatural entity.

To treat *obeah* pregnancy, the Q'eqchi' healer would have to fast from food and abstain from sexual activity. At midnight the family of the woman would have to throw a feast to appease the evil spirit that is inside the woman. The feast could perhaps be turkey. The healer, the woman with the *obeah* pregnancy, and her family would have to eat all of the food, the entire feast. The healer would then give the woman herbs to drink and call the evil being to be born. Sometimes only "wind" comes out, and a breeze is felt when it is born; other times some kind of animal is born such as a frog, snake, or scorpion. Sometimes a moving ball comes out that has nothing really much inside of it. Birth of the animal or spirit could be induced by prayer after a few months, not necessarily the term of a normal pregnancy. The healer would ask the woman to close her eyes after it is born, so as not to frighten her. Whatever form the evil spirit is born in, the healer must kill. He says that sometimes people come to him thinking they may have an *obeah* pregnancy, but it is not, it is just a normal pregnancy. He can tell through feeling the pulse. If a woman does have an *obeah* pregnancy she will feel terrible pain because the animal may be biting her insides. She will also experience headaches, nausea, and vomiting.

1.6.6 Hot–Cold Duality in Maya Medicine

The classification of medicines, foods, and illnesses as hot or cold is well documented in Latin America and the Caribbean, in many places including Oaxaca, Mexico (Messner 1981); western Guatemala (Tedlock 1987; Cosminsky 2001); northwest Argentina (Hilgert and Gil 2007); and Trinidad (Wong 1976). Cosminsky (2001) explains that in the hot–cold or humoral principle of Maya medicine, "hotness and coldness are considered innate or inherent characteristics possessed by the substance in question, although they may be influenced by physical temperature. Illnesses are caused by an imbalance in the body of excess cold or heat." Imbalances in the hot–cold equilibrium are treated with medicinal plants or treatments of the opposite humoral quality (Messner 1981), and traditions dictate which hot remedy would be used to treat which cold illness as not all hot remedies would be efficacious solely due to their humoral classification (Wong 1976).

In addition to illness, the humoral state of the body is influenced by the weather; the time of year; and time of day, age, and physiological state. Pregnancy and lactation, usually classified as hot physiological states (Hilgert and Gil 2007; Kay and Yoder 1987; Messner 1981). Among the Qiché-speaking Maya in Guatemala, menstruation, pregnancy, and birth were considered hot conditions because the body is thought to contain excess heat at those times, however after giving birth a woman is in a cold condition because of the loss of blood (Cosminsky 2001). Among the Yucatec Maya, 88 % of the medicinal plants used for health disorders specific to women, were classified as hot, suggesting also that the majority of women's health ailments are classified as cold ailments (Ankli et al. 1999). Spiritual illnesses on the other hand, such as "fright" or "evil eye" are typically considered hot conditions and treated with cooling remedies (Messner 1981). In many communities, such as among the Mitla Zapotec in Oaxaca, Mexico (Messner 1981), the classification is actually a hot–cold continuum, divided into states including "very hot," "warm," "temperate," "cool," "cold," and "very cold." Research on the hot–cold continuum has been criticized for reduction to a dichotomy with or without a central third term in the classification system (Tedlock 1987).

Q'eqchi' people classify physical and mental states, medicines, and foods as either "hot" or "cold," with a few "neutral" exceptions. Hot and cold are referred to as *chik i ke* in Q'eqchi'. This classification does not refer to the actual thermal state of the food, plant, or person. Rather this is more of a culturally attributed symbolic categorization, association, and value, and in some cases a personally attributed symbolic categorization. However, the classification may determine the thermal temperature of preparation or consumption of a particular remedy, treatment, or meal. In further studies on Q'eqchi' ethnomedicine it would be worthwhile to investigate the hot or cold classification of ailments themselves. This study focused on the plants and their properties and uses. The hot or cold classification of the remedy will often indicate the opposite classification of the disease. Sometimes the same diseases are treated with either hot or cold remedies and this may indicate dual forms or manifestations of diseases.

The plant and its preparation will be chosen to address the thermal disequilibrium of that disease. The remedy for internal heat or cold is to medicate with the plant with the opposite property. In other words, people with a hot illness would be treated with a cooling herbal remedy made from a cold plant. People with cold illnesses are treated with hot plants and warming remedies. Typically a hot remedy would be prepared hot, by boiling or pouring boiling water over the plants, and administered hot or warm, as in a tea or perhaps warm bath, and typically a cold remedy would be prepared cold, by crushing fresh leaves in cold water, and administered cold as in a cold drink, cold bath, or both. Most plants seem to be either hot or cold regardless of what different range of ailments it may be used to treat. Though there are exceptions like plants that are considered to be able to be made both into hot remedies for cold afflictions as well as cold remedies for hot afflictions. Other plant remedies may be made hot, by boiling and cooled before administered, and considered to be dual hot and cold plants simultaneously.

Ob'el (*Piper auritum* Kunth) leaves are mashed in water and a person bathes in the water to treat pain from "heat and cold." *Yut it* (*Piper peltatum* L.) leaf is used to treat pain anywhere on the body. The leaf is warmed, placed on the affected area, and covered. This remedy is good for the pain caused from cooking and then touching cold water, and good for *chik i ke* (heat and cold). *Yut it* is a "hot" medicine and used at night when cooking and going outside happens around the same time. *Ob'el* leaf can be used together with *yut it* for this. For children that cry too much because of too much "heat" in the body *saq q'ehen* (unidentified) would be used to make an herbal bath to bathe the child every morning for 3 days. This is also good for children and adults who are restless and cannot sleep because of too much "heat in the body."

Air and water temperatures are also classified as hot and cold, which may refer to their thermal qualities, and are believed to affect the body's thermal equilibrium. "Cold" can enter the body at the time of birth and during the postpartum period. An area in the house should be curtained off to warm the mother and new baby for 3 days, when there should be coals under the woman's bed and a warmed rock rubbed on her abdomen. After the birth, this should be done with the warmed rock every 20 or 30 min, and helps to ease postpartum pain and heal the uterus. Some say to then continue to do this once a day for the first postpartum month. After having a baby, the mother should not go into the winds (outside) for 1 week because it could enter her ear and she may not be able to hear. For 1 month the child should not be taken into the wind or out into the night.

Browner (1985) reported that in Cali, Colombia "menstruation, pregnancy, and parturition are times of greater than normal warmth and leave the woman particularly vulnerable to attacks of 'cold'…Women often describe sterility—the ultimate malfunctioning of the reproductive system—as a 'cold womb.' They believe that a delayed menstrual period may be due to a 'cold vagina.'" Similar beliefs have been documented in Asian societies, where cold wombs are considered barren and heat is considered to foster or restore fertility, and women are particularly vulnerable to the effects of "cold wind" during their postpartum period (Kendall 1987).

Women should not take a cold bath for a week, or others say a month, after giving birth. Women should not eat cold food for 1 month postpartum. A woman can consume a drink made with *black pepper* (*Piper nigrum* L.) to warm herself during the postpartum period. *Ob'el* (*Piper auritum* Kunth) is used as a postpartum treatment approximately 1 h after the birth of the baby and placenta. Three to five big leaves are boiled in a pot with about 1 gallon of water. This is then poured into a small bucket and the woman would sit on this until the water is barely warm anymore. This is done once a day for the first month. If this is done, preferably early in the morning, the woman should be able to get pregnant again easily. This brings out the "wind" and the "cold" and doing this for a month will bring the belly back down to normal size. If a woman cannot have a baby she is massaged several times and *pimienta gorda* (*Pimenta dioica* (L.) Merr.) is boiled with *cinnamon* (*Cinnamomum* sp.) stick and given to the woman to drink to "take out cold from inside."

Girls and women can also get “excess cold inside” due to drinking too much cold water or eating too much cold food during menstruation, or from soaking in too much cold water or bathing too long during menstruation. It is believed that not minding these practices can lead to excessive pain during menstruation, births, and menopause, and can lead to infertility. There is a belief that a menstruating woman should not hold a baby because that woman is too “hot” and the baby will get thin and wrinkled.

1.6.7 Maya Diagnostics

Diagnosis primarily involves listening to the patient’s symptoms and feeling the patient’s pulse. A traditional healer will feel the pulse of all his patients. Sometimes the healer will also ask if the patient has dreamed recently and then will interpret the dream to aid in the diagnosis. The most adept Q’eqchi’ healers can tell that patients have certain illnesses just by looking at them. One healer explained that some people just want to come and buy medicinal plants from him, but that he does not want to sell to them because he has not done the diagnosis himself. The patient’s pulse can be palpated at the ankles, wrists, or forehead. You can feel the pulse in any of those parts, so a healer may only feel one, or may feel the pulse at both the wrists and ankles for example.

“Each illness has a different way the blood flows in the body, so that’s how I know what is wrong with a person,” a healer explained. He feels the pulse along with praying a particular prayer to diagnose the patient and states that “with each illness the pulse feels different.” The pulse of a person with a physical ailment feels very different from the pulse of a person with a spiritual ailment; prayer is also used to aid in this diagnosis. One healer says when an ailment is spiritual or supernatural he would feel it in the patient’s pulse because the “vein pumps fast like how the heart beats fast when scared.” Others say the pulse beats slowly when someone is afflicted with a spiritual disease.

It is likely that different spiritual ailments have different pulses ranging from slow to fast, as in the well-defined pulse diagnosis system of Don Elijio Panti (Balick et al. 2008). Don Elijio, now deceased, was a revered and accomplished Maya healer who resided in the Cayo District of Belize. We found that he recognized at least 28 distinct pulse types and diagnosed 42 conditions that we know of by feeling his patients’ pulses. He said a child with very bad *susto* (“fright”) or an adult with a combination of hot and cold diseases would have an irregular, erratic pulse; an adult with *susto* has a fast pulse; cold in the body is marked by a subtle pulse; a pulse with long intervals between beats signifies “bad wind” disease; and a person with a rapid, fat, and jumpy pulse signifies that the patient suffers from a spiritual disease such as witchcraft or envy. Don Elijio used pulse palpation not only as a diagnostic tool, but also as a therapeutic tool by saying prayers “into the pulse,” and a means for tracking patients’ progress (Balick et al. 2008). Pulse diagnosis is also used to track patients’ progress in Q’eqchi’ traditional medicine.

1.6.8 Traditional Q'eqchi' Healing Modalities

The healers explain that medicinal plants taken internally alone cannot really heal the patient. The plants must accompany prayer, massage, and bathing with medicinal plants. Q'eqchi' treatments for reproductive health include: (1) teas made from medicinal plants, (2) vaginal steam baths containing herbs, (3) external bathing with water containing herbs, (4) abdominal and whole body massage, (5) prayer and ritual, as well as (6) recommendations for dietary and lifestyle changes. Suggested dietary and lifestyle modifications include recommendations such as no heavy lifting, banding the abdomen with a cloth, and no cold bathing or consumption of cold foods.

Abdominal massage is used for treating heavy menstruation, menstrual pain, hemorrhage, menopausal hot flashes, gastritis in men and women, stomach pain, *q'an yaj* (male pain around the navel), *k'uub'sa'* (womb disorder), female infertility, positioning the uterus, positioning a baby during pregnancy, female contraception reversal treatments, miscarriage prevention, turning breech babies in the womb, postpartum treatments, speeding childbirth, worms in children, "mole drop" (fallen fontanel) in children, and *susto* ("fright") by healers and herbalists. Special prayers are used for these in addition to the massage. For instance, Don Francisco explained you cannot just massage gastritis any way, or it can cause damage.

One herbalist knew a bit about Maya "acupuncture." Using a pine needle, she says three pricks in the forehead are used to treat migraines. The needle is stuck in a piece of dried *copal* (*Protium copal* (Schltdl. & Cham.) Engl.) resin to make the needling tool. She does not really practice it, but knows of someone who still does. Another woman says she does not do things in the same way as other healers, but begins with a prayer to God and specializes in massage therapy.

1.6.9 Medicinal Plant Harvest and Use

Medicinal plants are central to Q'eqchi' traditional medical treatments. For reproductive health, plants are used in herbal teas and vaginal herbal steam baths primarily. Herbal baths, poultices, tinctures, and applying fresh leaves topically are also used in Maya traditional medicine. The healers say they use the plants that are effective and only continue to use the ones they see results with. Before collecting plants, *ilonels* will adhere to certain practices that vary slightly from healer to healer. Don Francisco asks God for his blessing and asks that the plant work for a particular illness. He says this blessing is also important because you also do not know what spirits have been there before, so he needs to pray so that the plant is effective. Another healer, Don Manuel explained that "God left the plants for us to heal one another," and that he prays to God before collecting plants. "You have to ask God to collect the plant and mention for what purpose the plant is for. This prayer is part of a 13 part treatment prayer also. You need to collect at least 13 leaves of each plant. The human body also has 13 sections," according to Don Manuel.

Usually they will also abstain from sexual activity before collecting plants. It is best to collect medicinal plant after a few days of sexual abstinence, but when someone needs help the healer must collect the plants needed at the time they are needed. If a healer knows he will be collecting plants he should abstain from sexual activity for a few days before. Some healers say in general they try to abstain from sex in case a patient comes. One healer joked in a light-hearted way that his wife was upset with us (me and my interpreter) because we were collecting plants that day, since that meant he was not permitted to have sex with her the night before collecting.

Many medicinal plant species have become harder and harder to find, according to the healers, due to habitat loss. Don Francisco says the habitat loss is from natural disasters and slash-and-burn agriculture. Another added that this was due to expansion of the population and destruction of the forest. The *ilonels* of the QHA harvest plants sustainably and pay close attention to the health of the plant populations from which they collect. They will leave leaves on individual plants when collecting and change areas from which they collect if they notice the population getting smaller (Bourbonnais-Spear et al. 2006).

Don Francisco says that he has to travel farther to collect medicinal plants now and that sometimes he searches all day and barely find anything he is looking for. If he cannot find some plant he is looking for, he follows this practice, "in this world there are three classes of plants—primary forest, secondary forest, and home. If you cannot find in the first class go to the second, and if not go to the third class (home garden)." In an emergency he uses what he can find close, then goes out farther after to look for other plants. Don Manuel agreed saying that "some of the plants I use are very hard to find. I could go farther but due to the big hurricane and burning of the forest, this has destroyed where the medicinal plants are found." Hurricane Iris devastated southern Belize in October 2001, destroying homes, schools, agricultural fields, and forested areas. Another man explained that it is becoming harder and harder to find certain medicinal plants, because people chop them as they do not know what they are anymore. Some healers transplant medicinal plants to gardens for tending (Fig. 1.7).

Some healers feel there are best times for collecting medicinal plants. Don Francisco says there are best times based on the Maya calendar, to collect medicinal plants for particular treatments. Depending on the ailment being treated, there will be best days to collect, specific to that ailment or type of ailment. But in the case of an emergency, you have to collect whenever. The best time of day according to Don Francisco is early in the morning, when dew is on the plants because they have more medicinal content then. Most others agree that early morning is best. One explained that collecting medicinal plants in the morning is best, when the dew is still on the plant and the plants are cool because during midday the plants lose water from the leaves. Additionally, early in the morning is felt to be the best time to collect medicine because it will be its most effective since no one else has touched it. Early morning is thought to be the best time to collect something you want to plant, as well.

One female herbalist said that women should not collect medicinal plants while menstruating. My female interpreter carried *garlic* (*Allium sativum* L.) cloves

Fig. 1.7 Traditional healer, Don Domingo, collecting medicinal plants from the forest for transplanting to a home garden

tucked in her bra while we were collecting plants in the forest with a male traditional healer near Punta Gorda one day. She said it was because she was menstruating, and this was important to ward off "tigers" (jaguars) and other animals. Walking home, we each picked some *wild basil* (*Ocimum campechianum* Mill.) plants with seed to bring back to plant at our respective homes. She said also that if women collect plants while they are menstruating the plants will wilt quickly. A mile or so later we were home, and much to my surprise the wild basil plant she was carrying was drooped over her hand, very wilted, while mine looked as it did when I pulled it from the ground.

To find the plants the healers and herbalists need, their practices differ. Women often use plants growing around their house, or their neighborhood. Some will perhaps collect near the forest edge or in fields. Women, especially in the villages, will mostly wear skirts and flip flops, not the pants and boots really required for traveling deep into the "bush" for collecting forest growing species. Male healers may also cultivate some medicinal species at home, but will travel much father for collecting species found only deep in the forest. In the case of the Belize Indigenous Training Institute, the male healers have also established and tend to a medicinal plant forest garden, within walking distance from the Southern Highway. Due to these differences in collecting habits, women and men tend to use some different species of plants for medicinal treatments.

For minor cases, Don Manuel says he uses plants he can find around the neighborhood, but for serious cases, he needs to go to the mountain to collect the plants he uses in treatments. One male healer says he collects his plants walking back toward the hills from Punta Gorda, "way back in the bush." Another healer explained that he collects plants from the mountains, riverside, the village, and the "bush," but that plants are harder to find here than in Guatemala where he used to practice as a healer before coming to Belize.

One woman says regarding the plants she uses that she "grows some here at the house, some are common[ly found], and some my husband has to find in the bush." Most women say they get their medicinal plants around the house, and others added they get plants from walking around while handling other chores and errands, or on visits to other villages to see family such as parents or grandparents. Some women send their husbands to the bush, or go together with their husbands, and sometimes women will go to the bush themselves. One older, traditional woman explained that she gets her plants from neighbors or she plants them herself, and that she does not go deep into the forest, but does gather from the "bush" near her house. She said that she, as a traditional Q'eqchi' woman uses all the plants she has around to treat sickness in the family. She collected all the plants for me with her own hands during our collection because the "heat" from my hands or body could dry out the plants or cause them to die, she said. She does not even let her own children collect her plants, and honors the tradition of collecting her medicinal plants with her hands.

Conclusions

The introductory material has provided the lens with which to read the richly detailed knowledge regarding Q'eqchi' reproductive ethnomedicine. Plant use is vital to Maya culture, and this study only highlights one of many areas of traditional plant-based knowledge. Scholars from a biomedical or scientific background should read with an open mind and consider that this system of traditional Maya medicine has maintained the health and fertility of the Maya people since ancient times. Maya medical concepts, diseases, and treatments may not all have correlates in biomedicine, yet there is much to learn in the wisdom the Maya have developed and honed over the last 4000 years living in the richly biodiverse Mesoamerican ecosystems.

References

American Psychiatric Association (2002) DSM-IV-TR: diagnostic and statistical manual of mental disorders, 4th edn. American Psychiatric Association, Washington

Ankli A, Sticher O, Heinrich M (1999) Yucatec Maya medicinal plants versus nonmedicinal plants: indigenous characterization and selection. Hum Ecol 27(4):557–580

Arvigo R, Balick M (1993) Rainforest remedies: 100 healing herbs of Belize, 2nd edn. Lotus, Twin Lakes

Balick MJ, O'Brien H (2004) Ethnobotanical and floristic research in Belize: accomplishments, challenges and lessons learned. Ethnobotany Res Appl 2:77–88

Balick MJ, Nee MH, Atha DE (2000) Checklist of the vascular plants of Belize. NYBG Press, New York

Balick MJ, De Gezelle JM, Arvigo R (2008) Feeling the pulse in Maya medicine: an endangered traditional tool for diagnosis, therapy and tracking patients' progress. Explore (NY) 4(2):113–119

Bourbonnais-Spear N, Poissant J, Cal V, Arnason JT (2006) Culturally important plants from southern Belize: domestication by Q'eqchi' Maya healers and conservation. Ambio 35(3):138–140

Browner CH (1985) Traditional techniques for diagnosis, treatment, and control of pregnancy in Cali, Colombia. In: Newman LF (ed) Women's medicine: a cross-cultural study of indigenous fertility regulation. Rutgers University Press, New Brunswick

Collins DA, Liukkonen JR (2002) What's in a name?: plant lexical variation among the Q'eqchi' Maya, Alta Verapaz, Guatemala. In: Stepp JR, Wyndham FS, Zarger RK (eds) Ethnobiology and biocultural diversity. International Society of Ethnobiology. University of Georgia Press, Athens

Cosminsky S (2001) Midwives and menstrual regulation: a Guatemalan case study. In: Van de Walle E, Renne EP (eds) Regulating menstruation: beliefs, practices, interpretations. The University of Chicago Press, Chicago

Crooks DL (1997) Biocultural factors in school achievement for Mopan children in Belize. Am Anthropol 99(3):586–601

Hartshorn G et al (1984) Belize country environmental profile: a field study. Robert Nicolait & Associates, Belize City

Hilgert NI, Gil GE (2007) Reproductive medicine in northwest Argentina: traditional and institutional systems. J Ethnobiol Ethnomed 3:19

Jacobs ND, Castañeda A (1998) Belize national biodiversity strategy. National Biodiversity Committee, Ministry of Natural Resources and the Environment, Belmopan

Jordan B (1984) External cephalic version as an alternative to breech delivery and cesarean section. Soc Sci Med 18(8):637–651

Kay M, Yoder M (1987) Hot and cold in women's ethnotherapeutics: the American–Mexican west. Soc Sci Med 25(4):347–355

Kendall L (1987) Cold wombs in balmy Honolulu: ethnogynecology among Korean immigrants. Soc Sci Med 25(4):367–376

Messner E (1981) Hot–cold classification: theoretical and practical implications of a Mexican study. Soc Sci Med 15B:133–145

Michel J, Duarte RE, Bolton JL, Huang Y, Caceres A, Veliz M, Soejarto DD, Mahady GB (2007) Medical potential of plants used by the Q'eqchi' Maya of Livingston, Guatemala for the treatment of women's health complaints. J Ethnopharmacol 114:92–101

Pesek T, Cal V, Fini N, Cal M, Rojas M, Sanchez P, Poveda L, Collins S, Knight K, Arnason J (2007) Itzama: revival of traditional healing by the Q'eqchi' Maya of southern Belize, Central America. Herbal Gram 76:34–43

Staiano KV (1981) Alternative therapeutic systems in Belize: a semiotic framework. Soc Sci Med 15B:317–332

Statistical Institute of Belize (2007) Abstract of statistics. Belmopan, Belize

Statistical Institute of Belize (2011) Main results of 2010 population and housing census. Belmopan, Belize. Accessed 24 Apr 2012. http://www.sib.org.bz/Portals/0/docs/publications/census/2010_Census_Report.pdf

Tedlock B (1987) An interpretive solution to the problem of humoral medicine in Latin America. Soc Sci Med 24(12):1069–1083

Weller SC, Baer RD, Garcia de Alba Garcia J, Glazer M, Trotter R, Pachter L, Klein RE (2002) Regional variation in Latino descriptions of susto. Cult Med Psychiatry 26:449–472

Wilk RR (1997) Household ecology: economic change and domestic life among the Kekchi Maya in Belize. Northern Illinois University Press, DeKalb

Wilson R (1993) Anchored communities: identity and history of the Maya-Q'eqchi'. Man 28(1):121–138

Wong W (1976) Some folk medicinal plants from Trinidad. Econ Bot 30(2):103–142

Woods LA, Perry JM, Steagall JW (1997) The composition and distribution of ethnic groups in Belize: immigration and emigration patters, 1980–1991. Lat Am Res Rev 32(3):63–88

Chapter 2
Traditional Q'eqchi' Maya Reproductive Ethnomedicine

Q'eqchi' Maya traditional medicine utilizes an extensive plant-based ethnopharmacopoeia for reproductive health and fertility regulation. This branch of traditional Q'eqchi' medicine also includes traditional practices and beliefs related to child bearing and gender-specific health. The goal of my ethnomedical research was to document, as comprehensively as possible, the plant-based remedies, traditional medical treatments, beliefs and practices related to reproductive health, in collaboration with the Belizean Q'eqchi' community. My collaborators were the male healers from the Q'eqchi' Healers Association (QHA), the few remaining Q'eqchi' midwives, and other women knowledgeable in medicinal plants, whom I refer to as "herbalists", living in the Toledo District of southern Belize.

There has been an accelerated loss of knowledge in women's healing traditions among the Q'eqchi' Maya living in Belize due to a number of cultural factors. Treating women's reproductive health was traditionally the realm of female healers; however, there are very few Q'eqchi' midwives who practice in Belize today, as traditional midwifery has been discouraged by the Belize Ministry of Health. Migration and loss of familial networks have contributed to the loss of women's healing traditions in Belize as well. Chapter 4 takes a closer look at these factors. This accelerated loss of a particular branch of traditional knowledge highlighted the critical need for a comprehensive study of Q'eqchi' reproductive ethnomedicine in Belize. What follows is a brief review of related research, the study's methodology, and the totality of the results from our collaborative ethnomedical study on reproductive health and fertility regulation.

Research on the ethnobotany of the Q'eqchi' Maya of Belize has included other collaborative studies with the QHA, including consensus ethnobotany of the QHA's use of medicinal plants and the diseases treated (Amiguet et al. 2005), a regression analysis of medicinal plants used by the QHA (Amiguet et al. 2006), and a rapid ethnobotanical survey of the Maya Mountains Range with the QHA (Pesek et al. 2006). A survey of medicinal plants used by the QHA to treat neurological and mental health disorders (Bourbonnais-Spear et al. 2005), and pharmacological research on plants used to treat the culture-bound syndrome, *susto*, for their anxiolytic effects (Bourbonnais-Spear et al. 2007) have been conducted. The QHA's cultural healing center and medicinal garden, Itzama, which means "home of the Maya god of

J. De Gezelle, *Q'eqchi' Maya Reproductive Ethnomedicine*,
SpringerBriefs in Plant Science, DOI 10.1007/978-3-319-10744-8_2

wisdom, Itzamna" and a place of spiritual and herbal healing, has been proposed as a model to be used internationally for holistic, sustainable healthcare in indigenous communities, and conservation of the associated biodiversity (Rojas et al. 2010).

While my study constitutes the only known research on plants used by the Q'eqchi' Maya specifically for reproductive health in Belize, a similar study was carried out in the Q'eqchi' communities surrounding the Livingston area of Guatemala, just one h across the sea by boat from Punta Gorda, Belize. Joanna Michel's doctoral dissertation titled *Medical Ethnobotany of the Q'eqchi Maya: Perceptions and Botanical Treatments Related to Women's Health* (2006) documented medicinal plants used for reproductive health, as well as cultural conceptions of and choices related to Q'eqchi' women's health in Guatemala. Nineteen medicinal species from Michel's study were screened for their ability to bind to estrogen, progesterone, and serotonin receptors, which is discussed in more detail in Chapter 3. Similarities to Michel's findings as well as other studies throughout Latin America and the Caribbean are mentioned throughout this chapter.

2.1 Methodology

2.1.1 Study Site and Timeline

My fieldwork in Belize began in 2006 in the western Cayo District, working with Yucatec Maya healers with guidance from Dr. Rosita Arvigo and Dr. Michael Balick. In 2007, I initiated research in the southernmost Toledo District, working with the Q'eqchi' Maya and in 2008 also began research with the Garifuna communities in Toledo as well. During the years of 2007–2011, I spent nearly 30 months doing fieldwork in Toledo, based in Punta Gorda Town. I interviewed traditional healers, midwives, and herbalists in Punta Gorda's Indianville neighborhood and the villages of Laguna, Indian Creek, Big Falls, Midway, and San Pedro Columbia. I made plant collections in forested areas surrounding Punta Gorda and the villages of Jalacte, Indian Creek, Laguna, and San Pedro Columbia.

2.1.2 Research Ethics

Research methodologies, including interview questions (Appendix I) and consent forms were reviewed and approved by The City University of New York's Institutional Review Board (IRB Number: 09-12-1892). The Belize Forest Department approved my research proposal and granted me a research and collection permit for the Toledo, Stann Creek, and Cayo Districts, though plant collections were only made in the Toledo District. A research agreement was signed between the researcher and the Belize Indigenous Training Institute (BITI). All interviews were undertaken only after obtaining prior informed consent, using a consent form provided in the participant's native language. Interviews were conducted in the participant's language of choice, using an interpreter when necessary.

Fig. 2.1 Walking interview with Q'eqchi' *ilonel*, Don Domingo and my Q'eqchi' Mayan interpreter and research assistant, Maria

Fig. 2.2 Photo elicitation interview with Ms. Francisca

2.1.3 Interviews

My approach to fieldwork was multidimensional, using several techniques that contributed layers of data that deepened, clarified, reaffirmed, and on the rare occasion,contradicted the larger body of my ethnographic findings. Semi-structured interviews, forest and home garden "walking" interviews (Fig. 2.1), photo elicitation (Fig. 2.2), plant collections, and observations of remedy preparations and treatments were used to collect data with 21 Q'eqchi' individuals (6 men and 15 women) knowledgeable in medicinal plants. These and other interviewing methodologies are discussed in detail in the work of Miguel Alexiades (1996). The 6 males interviewed are recognized healers, but only 3 of the 15 women are recognized as traditional healers, specifically—midwives. One of those midwives is no longer practicing, and one agreed to do the interview with my interpreter only, so I did not have the opportunity to meet her. The third midwife, who was still practicing, had moved to Belize from Guatemala only in recent years.

The Q'eqchi' traditional healers, or *ilonels*, I interviewed were identified through the BITI, under which operates the QHA, at that time an association of six male *ilonels*. From there my two interpreters suggested a few midwives and many knowledgeable women, whom I refer to as "herbalists," that they knew of. Beyond this, my interpreters inquired in their networks for other knowledgeable people their family and friends knew of.

Semi-structured interviews lasted 45 min–3 h and typically took place at the home of the person being interviewed, either inside the house or outside near the house. Occasionally, the interviews were held at my interpreter's house or at the house I was renting in Punta Gorda. Interviews focused on plants used to treat reproductive ailments, in particular conditions sometimes linked to estrogen deficiency, such as infertility, menopausal complaints, and lack of menstruation, as well as for conditions related to estrogen dominance in the body, such as heavy or painful menstruation. Detailed information was gathered regarding the procedures for collection, preparation, and administration of remedies. Other sections of the interview covered socio-demographic data, life history, aspects of their healing practice, and beliefs regarding reproduction and reproductive health.

2.1.4 Plant Collection and Identification

I made plant collections in forested areas surrounding Punta Gorda and the villages of Jalacte, Indian Creek, Laguna, and San Pedro Columbia with the healers I interviewed. Trips were made to collect voucher specimens for the species discussed in our interviews, as well as bulk samples for the species selected for bioactivity screening. Plants were collected at sites where the healers collect plants or other sites that were more accessible where they know the species grow. Plant species were identified with the aid of the *Checklist of the Vascular Plants of Belize* (Balick et al. 2000) and the collections at The New York Botanical Garden's herbarium (NY). Daniel Atha and Ricardo Kriebel from The New York Botanical Garden made many of the final determinations of the specimens collected in Belize for this study. When possible, collections were made in triplicate. One set of specimens is deposited at The New York Botanical Garden herbarium (NY), one set at the Belize National Herbarium (BRH), and one set with the BITI.

2.2 Results and Discussion

Traditional remedies; treatments; beliefs; and practices regarding reproductive health, fertility regulation, and infant care are reported here. Plant names used by the healers and herbalists are in italics text, with additional local names parenthetically listed preceding the Latin names and authors of the botanical nomenclature. Only four Q'eqchi' research participants are referred to by name: Don Francisco and Don Manuel of the QHA and

Ms. Francisca and Ms. Adelina, who are elder Q'eqchi' herbalists. These two men and two women were among my closest collaborators.

Don is the Spanish prefix used to denote respect for a male elder, and is utilized when referring to male *ilonels*. Don Francisco is considered to be the most knowledgeable *ilonel* or traditional healer involved with the BITI. He specializes in reproductive health among a few other areas of traditional medicine, but treats nearly all health conditions. Don Manuel is an elder healer, also very knowledgeable in treating all types of health conditions. He has an impressive home garden with a vast variety of edible and medicinal plants, and built his own traditional wooden violin that he would play for us after interviews.

The female equivalent of *Don* is *Doña*; however this is not commonly used in the Belizean Q'eqchi' community. In English, Belize's national language, any older woman is referred to as Ms. (though pronounced "Miss") followed by her first name, out of respect. So, herein I refer to the traditional healers and herbalists how I address them. Ms. Francisca's family addresses her as *Na'chin*, which means Grandmother in Mopan Mayan, which much of the family speaks in addition to Q'eqchi' and English. Ms. Francisca is a knowledgeable herbalist, a woman who has been using medicinal plants to treat herself, her children, grandchildren, great grandchildren, and close friends and neighbors for decades. Ms. Adelina is a farmer and a knowledgeable herbalist who maintains many traditional practices in her plant cultivation and daily life. She has her own *milpa* where she grows an impressive variety of fruits and vegetables that she brings to Punta Gorda from her village to sell on market days. They are both elders in their community and familial networks, and are each a wealth of Q'eqchi' women's traditional knowledge.

2.2.1 Disorders of the Womb

In Q'eqchi' Maya ethnomedicine, there is a culturally specific uterine condition, known as *k'uub'sa'* ("disorder of the womb") or *ke sa'* ("cold inside") which can be caused by displacement of the womb, or "cold" entering the womb. In Spanish, it is known as *matriz* or *enfermadad de matriz* ("womb" or "womb illness"). According to Don Francisco, the age range of women affected by *k'uub'sa'* is 14–48; however it is most common in women aged 20–30. He says some of the cases he has treated were due to lack of postpartum abdominal massage "when the womb is typically open," lifting heavy objects, falling during menstruation, having sex too soon following childbirth, and physical abuse. Others added that this disorder can also be caused by poor sleeping positions, falling by the river or anywhere anytime though especially during menstruation, heavy lifting during menstruation, or not binding the abdomen with a cloth after childbirth.

Don Francisco says he knows when a patient has this condition because the woman's menses will be too heavy, come too early, or will be otherwise abnormal, along with lower back pain and abdominal pain. He also says you can feel *k'uub'sa'* in the pulse because it will feel slower than normal. According to another traditional healer, symptoms of "cold in the womb" include excess pain in the

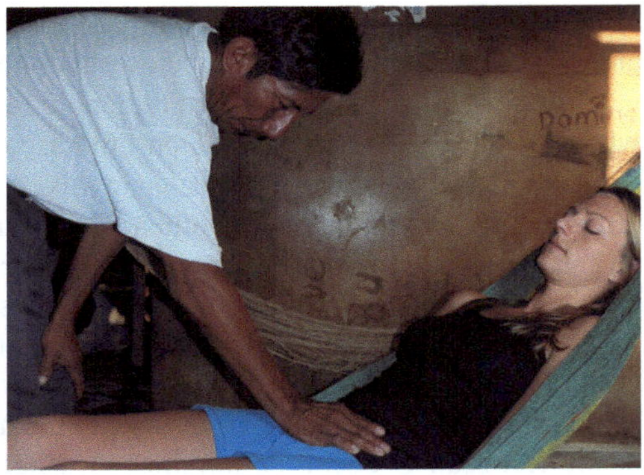

Fig. 2.3 Don Francisco demonstrating his womb massage with naturopath Dr. Zoe Palmer-Wright, beginning with the prayer for this treatment

stomach, crunching sounds when the stomach is pressed, and abnormal stool. *K'uub'sa'* is treated with medicinal plants coupled with prayer, massage, and "feet knocking" while the patient is lying in a hammock. A healer will "knock" or gently hit the bottom of the woman's feet that are sticking high up in one end of the hammock she is laying in. Using only the medicinal plants by themselves is not sufficient Don Francisco says, but the whole treatment is very effective. "Knocking the woman's feet is part of the secret to putting the womb back into place, since it has fallen down, the knocking with the feet up high in the hammock will put the womb back up," he explained. Another healer said they "knock the feet because it is tradition and knocking the feet three times will set the womb back in place."

Don Francisco would give three massages, with 3 days in between each massage. He says the prayer he uses while giving the massage is more important than the touch of the massage in treating *k'uub'sa'* (see Fig. 2.3). The most important part is asking God to help him set the womb. The massage and prayers last 10–15 min. The massage involves long, light strokes from the ends of the legs toward the center of the woman's abdomen, and in upward motions from the lower to middle abdomen.

Don Francisco uses a plant called *kux sawi'* (*Piper tuerckheimii* C. DC. ex Donn. Sm; JD 1) to treat *k'uub'sa'* or the associated heavy menstruation. The exact origin and meaning of the Q'eqchi' name is unknown; however, *Kux* is the Q'eqchi' version of the name *Marcos* and *Sawi'* could have been a surname. Don Francisco is not sure, but says perhaps this is the origin of the plant's Q'eqchi' name, and that perhaps it references a healer who may have used and taught the remedy. To treat *k'uub'sa'* or heavy menstruation, he would uproot two whole plants and boil them in a quart of water for 1 h. The whole plant is used, including the root, stem, and leaves. The woman should drink a glass in the morning and one in the evening until she gets better. He says this treatment works very effectively because he first performs a healing including abdominal massage to "fix the womb within the woman" and pulls her feet up high in the hammock and "knocks" her feet to restore her womb. It is very effective in 2 or 3 days he says. According to Don Francisco, this plant could be dried and stored and be effective for 1 or 2 months after collection. He has to go far out into "the bush" now to collect the plant and has difficulty finding it because of the damage to the forest in the last 20–25 years caused by hurricanes and slash-and-burn agriculture. He says the plant is found growing mostly in primary forests, close to the hills or riversides. It is considered a hot plant in his system of classification. Another traditional healer also uses *kux sawi'* to treat women who are experiencing problems with the womb, such as womb pain. He boils a handful of the leaves, and the woman drinks one cup of the tea, three times a day until she gets better.

To treat a woman with *k'uub'sa'*, Don Francisco could also use *yut it* (*Piper peltatum* L.; JD 15). The name of the plant is translated to "tied bottom" and is named so because all the leaf veins join together to one stem (the plant has peltate leaves). Don Francisco uses the whole plant including root, stem, and leaves. To treat *k'uub'sa'* he would boil the whole plant, and the woman would drink a warm cup twice a day, once in the morning and once in the evening until she gets better. This treatment would also be accompanied by massage to put the womb back in place. Massage, prayer, and "feet knocking" are all done before the plant remedy is given. This plant is considered a cold remedy according to Don Francisco's classification system. The root and stem of this plant can be stored for later use, but not the leaf. The plant is common and found only in secondary forests he says.

For treating "cold in the womb" caused by not bathing in warm water during menstruation, one healer uses *puchuch q'ehen* (unidentified; JD 85). He says there are two plants with the same name, but one has darker leaves and the other has lighter leaves. He would boil the darker-leaved one with the roots of *yut it* (*Piper peltatum* L.; JD 15) and the "skin" or bark of *saq xook'* (unidentified). The patient would drink one warm cup. According to this healer, not many people know this remedy.

For a fallen womb caused by falling and lifting heavy things, the woman would be treated by having her feet "knocked," her abdomen massaged upwards, and then banded with a cloth and kept on for 1 week. A shifted womb can cause excessive menses that is treated with massaging as well as binding of the pelvic region. Women are not allowed to do heavy lifting during this treatment. One midwife said she would also boil five open *hibiscus* (*Hibiscus rosa-sinensis* L. var. *rosa-sinensis*) flowers and four closed *hibiscus* flowers and have the woman drink this tea three times a day during her period. Her husband, a traditional healer, said he would treat a woman whose womb is out of place with prayers, medicinal plants, hammock work, and words to bind the abdomen, not a cloth, he emphasized.

If a woman has "fibers in the womb" (uterine fibroids), a healer explained that "the woman's belly will feel round, even though she is not pregnant, and she will have excess pain in her belly." To treat this he would use white rum with whole *black pepper* (*Piper nigrum* L.), *cinnamon* (*Cinnamomum* sp.), *cloves* (*Syzygium aromaticum* (L.) Merrill & Perry), *anise* (*Pimpinella anisum* L.), and honey together in a bottle, which is then buried for 24 h. He would then have the woman drink this regularly. *Pimpinella anisum* L. is also used by Latino healers in New York City to treat uterine fibroids (Balick et al. 2000). This healer said he could also collect *wara k'ix* (sleeping prickle: *Mimosa pudica* L.; JD 50), *saq pach'aya'* (unidentified; translated as "white grass"; JD 89), and the root of *kwoyo'* (possibly *Triumfetta semitriloba* Jacq.) to treat "fibers in the womb." The three are boiled together and given to the woman to drink in the amount of 1 gallon/day.

2.2.2 Menstruation

Ms. Francisca believes that women who are menstruating should not take any cold baths or have any cold food or they will have excess pain during their periods, births, and menopause. According to Don Francisco, women are not supposed to bathe for long periods of time during their menstruation nor in water that is too cold. The Qiché Maya in Guatemala also feel that a woman should not bathe in cold water when she is menstruating, as the hot/cold extreme of her hot menstruating state and the cold water could cause her menstruation to cease and her abdomen to swell (Cosminsky 2001). The Q'eqchi' in Belize also feel that women should not eat or drink things that are too cold during menstruation and should also take care not to over work or lift heavy things. There is a Q'eqchi' belief that a menstruating woman should not hold a baby because menstruating women are too "hot" and the baby would get thin and wrinkled, as a result. Similarly, among the Qiché in Guatemala, menstruating and pregnant women are felt to be "hot" and can cause illness to babies unintentionally or intentionally through their admiration of others' children, in the form of the "evil eye" caused by their excess heat (Cosminsky 2001).

In the Dominican Republic, "women who are menstruating are not allowed in *conucos* (small plots of cultivated land) or to touch flowers because their presence will harm plants. In addition, they are not allowed to collect or touch plants because it may cause the plants to wither and to have spots or blemishes. Women who are menstruating are neither allowed to touch babies or carry them, nor can they cut a person's hair. Although many still believe these taboos, there are many who no longer do" (Ososki 2004).

My female interpreter carried *garlic* (*Allium sativum* L.) cloves tucked in her bra while we were collecting plants in the forest with a traditional healer one day. She said it was because she was menstruating, and this was important to ward off "tigers" (jaguars) and other animals in the forest. Walking home, we each picked some *wild basil* (*Ocimum campechianum* Mill.) plants with seed to bring back to grow at our respective rented houses. She said that if women collect plants while they are menstruating, the plants will wilt quickly. A mile or so later we were home, and much to my surprise the plant she was carrying was drooped over in her hand, very wilted, while mine still looked just as it had when I pulled it from the ground.

Dysmenorrhea (Painful Menstruation)

Painful menstruation is due to "too much cold inside" and is treated with prayer, rubbing a warm rock over the abdomen, and teas of medicinal plants. Don Francisco massages his patient to treat for dysmenorrhea, by putting her in a hammock with her feet higher, and her head lower than normal in the hammock. He would "knock" her in the feet three times and continue with the massage, including massaging the womb, as he would for *k'uub'sa'*, which is the likely cause for the dysmenorrhea.

Don Francisco uses *chinta pim* (1) (*Desmodium incanum* DC.; JD 9) to treat back pain associated with menstruation. The name comes from the Spanish word *cintura*, which means a belt or ribbon used to tie or hold, since the seeds stick to your clothing when you brush past the plants. *Pim* means herb; in this case, medicinal plant. Don Francisco uproots the whole plant to use. He says the plant is considered either a hot or cold remedy because it is boiled, but then it is cooled down to be used. Some plants are considered hot or cold according to their uses, as in the temperature of their preparation and administration. This one is like a 50/50 usage he says. He boils the plant and leaves it to cool down. Then the woman would drink some of the water, and bathe in the rest. She would drink it once a day every morning for 3 days and bathe every morning in it as well. He says this is a common plant that grows in primary and secondary forests. The plant could be stored for 2–3 months. Another species in the same genus, also referred to as *chinta pim* (2) (*Desmodium axillare* DC.; JD 11), is used for the same purpose. Both are prepared and administered in the same way for back pain associated with menstruation.

Yut it or *yut u it* (*Piper peltatum* L.; JD 15) leaves are wrapped around a warm rock and rubbed on the stomach for 2 days, in the morning and in the evening to treat menstrual pain. *The Handbook of Kekchi Medicinal Plants of Belize* (Maquin et al. 2005) reports that the leaves are crushed in cold or hot water and drunk to treat a heavy menstrual cycle and the liquid is used to wet the area of the body with associated menstrual pain. According to one healer, if someone does not want to use *yut it* they can use *copal* (*Protium copal* (Schltdl. & Cham.) Engl.) and rub that on their stomach instead. Other plants mentioned for treating menstrual pain were *kok'mox* (*Maranta arundinacea* L.; JD 6), *b'aknel pim* (*Drymonia serrulata* (Jacq.) Mart.; JD 16) and *kux sawi'* (*Piper tuerckheimii* C. DC. ex Donn. Sm; JD 1), along with prayer. One traditional healer uses *rix choql* (unidentified), *saq q'ehen* (unidentified; JD 48), and *saq kampaan* (unidentified) leaves boiled together to ease menstrual pain, as he says this remedy "can bring down the cold."

For severe menstrual cramps, one midwife suggested holding pressure points between the thumb and forefinger. The young leaves of the *achiote* (*Bixa orellana* L.) tree can be made into a tea to drink three times a day for treating cramps. The water is boiled and then the young leaves are mashed into the water. For painful periods, some women use *serosi* (*Momordica charantia* L.; JD 69). A small handful of the vine and leaves are boiled in three cups of water, until the water turns dark and yellowish. Three cups of tea are consumed warm daily before meals. Q'eqchi' women in Livingston, Guatemala also use *Momordica charantia* L. leaf tea to treat dysmenorrhea (Michel 2006).

In a remedy for menstrual cramps and "cold inside," the leaves of *yerba cancer* (*Acalypha arvensis* Poepp.; JD 54) and *albahaca* (*Ocimum campechianum* Mill.) are boiled together and taken as a tea. One entire *b'enk* (basil: *Ocimum campechianum* Mill.) plant can be boiled in approximately four cups of water for 5–10 min, for menstrual cramps, which is consumed three times during the day. *Frijoli* (thought to be *Senna occidentalis* (L.) Link) is used for painful menstruation. The roots are boiled in a small amount of water for 5–10 min. This is consumed three times a day during the menstrual cycle. For menstrual pain, six "*pear*" (avocado: *Persea americana* Mill.) leaves are boiled and the woman would drink one cup of the liquid in the morning and one cup in the evening. *Pimienta gorda* (*Pimenta dioica* (L.) Merr.)

Fig. 2.4 *Yerba de cancer* (*Acalypha arvensis* Poepp.) leaves and flowers

seed can also be boiled as a tea to use for alleviating menstrual cramps. The fruit of *P. dioica* is also used for the treatment of menstrual cramps in the Petén region of Guatemala, boiled along with avocado seeds, cinnamon, and cloves (Comerford 1996). *Pimenta dioica* leaves are used for managing menopausal symptoms in Costa Rica, and have been shown to be a partial agonist/antagonist of human estrogen receptors (Doyle et al. 2009).

Yerba de cancer (*Acalypha arvensis* Poepp.; JD 54; Fig. 2.4) is used to treat painful menstruation with odor, indicating an infection, according to one herbalist. She would use one handful of the plant to make one cup, which would be given to the woman to drink two or three times a day for 3 days. For vaginal infection, she says a woman should take a bath then wash the vagina with a small bowl of water containing three drops of vinegar with half a lime (*Citrus aurantifolia* (Christm.) Swingle). *Chupil pim* (unidentified) is used for menstrual cramps, pain, excess period and menstrual odor. Approximately three entire plants are boiled with one clove of garlic (*Allium sativum* L.) and black pepper (*Piper nigrum* L.) seed for 20 min. The woman would drink this approximately three times per day during her menstruation.

Menorrhagia (Heavy Menstruation)

Don Francisco uses *ch'ajom k'aham* (*Gouania lupuloides* (L.) Urb.; JD 4) to treat heavy menstruation, as a "stoppage for the heavy flow of blood." *Ch'ajom* means young man or boy, and *k'aham* means vine. Either the bark or the leaves could be used for a tea. He would collect a 6-in piece of vine "bark" scraped off the vine, cut to pieces and boiled, or 12 or 20 leaves boiled to make tea for the woman to drink. A woman would drink this twice per day, once in the morning, once in the evening for 2 days or until she gets better. This is a hot remedy, according to his classification system. This is a common plant found in primary or secondary forests, though it takes a long time for it to grow into a big vine, according to Don Francisco. He says the bark could be stored for over a year and could still be used. *Gouania lupuloides* (L.) Urb. was reported to be used by Latino healers in New York City for the treatment of uterine fibroids (Balick et al. 2000), a condition that typically causes heavy menstruation.

The significance of using 12, 13, or 20 leaves in his practice rests in the calendar, he says. In the Tzolk'in Maya calendar, there are 13 cycles, each containing the same 20 named days with their 20 glyphs and particular attributes associated with the days. The specific number of leaves Don Francisco uses in remedies seems to be a syncretic practice, using the number 12 as it correlates to the months in the standard Belizean calendar sometimes instead of the 13 cycles of the Maya calendar; whereas the amount of 20 he explicitly links to the Tzolk'in Maya calendar cycles of 20 days. The numbers 12, 13, and 20 come up repeatedly in his remedies.

Don Francisco also uses *ichaj pim* (*Psychotria acuminata* Benth.; JD 5) for the same purpose as *ch'ajom k'aham* (*Gouania lupuloides* (L.) Urb.; JD 4) as "stoppage for heavy menstrual periods." *Ichaj* is the word for *chaya* (*Cnidoscolus chayamansa* McVaugh; also called *maak'uy*), which is similar in appearance to this plant, so that is how the plant got its name. *Pim* means herb. He uses 20 of the *ichaj pim* leaves crushed in cold water for the remedy. This would be given to the woman to drink three times during the day for 1 day only. It should work after 1 day. This plant could be stored but probably only for a month or two since it is a soft plant, according to Don Francisco. This plant is considered a cold plant according to his system of classification. He says it is a common plant and can be found in primary or secondary forests.

Regarding which is better for stopping heavy menstrual flow, *ichaj pim* or *ch'ajom k'aham* he said that one is cold and one is hot. So if she is heated in her stomach she requires something to cool down her body, which would be *ichaj pim*. *Ch'ajom k'aham* is used to heat a woman up when she has cold inside. It depends on the woman and whether she needs to be cooled down or heated up; this will determine which plant remedy Don Francisco uses in this case.

To treat heavy menstruation (and postpartum hemorrhaging), Don Manuel also uses *ichaj pim* (*hikalit kak*: *Psychotria acuminata* Benth.; JD 5) leaves. He noted that the plant also has the name, *hikalit kak* because of the purple undersides of the leaves. A handful of leaves are boiled and the woman drinks just a cup, if it works right away. She would drink more if she is still bleeding. In this case, she would drink the tea three times a day until it works. This plant is considered a hot plant according to Don Manuel's classification system. The plant is not common, but is found in both primary and secondary forests, he says.

Don Francisco says the whole plant of *kux sawi'* (*Piper tuerckheimii* C. DC. ex Donn. Sm; JD 1), including leaves, stems, and roots, can be boiled and given to the patient to drink and bathe in to treat heavy menstruation. *The Handbook of Kekchi Medicinal Plants of Belize* (Maquin et al. 2005) reports that *Piper tuerckheimii* C. DC. ex Donn. Sm. is also used to treat postpartum bleeding by giving the woman a decoction of the leaves to drink warm.

Don Francisco also uses *yut it puchuch* (unidentified), which is found mostly in primary forest to treat heavy menstruation. He collects many of the roots and some leaves and boils for the woman to drink. To treat heavy menstruation, one midwife also said she would boil *yut it puchuch* (unidentified). The tea would be consumed warm three times a day for 3 days. Three or four leaves boiled in a quart of water are enough for 1 day. Women are not to bathe in or drink cold water during this treatment because they will swell according to her. *Kok' mox* (*Maranta arundinacea*

L.; JD 6) and *yut pim* (*tyut it: Piper peltatum* L.; JD 15) were also mentioned as remedies for heavy menstruation.

One herbalist says to boil and drink *b'enk* (basil: *Ocimum campechianum* Mill.) roots to treat heavy menstruation. Other remedies for heavy menstruation use the leaves and call for boiling five *b'enk* leaves to make a cup of tea, or a handful of *b'enk* leaves and drinking the tea twice, or making a tea of *pom pim* (*Piper* sp.) and *b'enk* leaves. For heavy periods one herbalist would cut up "one dollar" amounts (packages of roughly two teaspoons of dried herbs, sold by the *cobaneros*—Guatemalan vendors who have stalls selling a variety of household items in the center of Punta Gorda town) of each of: *manzanilla* (thought to be *Matricaria recutita* L.) stems with leaves and flowers, *oregano* (thought to be *Lippia graveolens* H.B.K.) leaves, *chalche* (unidentified) leaves, along with *garlic* (*Allium sativum* L.) and three *cloves* (*Syzygium aromaticum* (L.) Merrill & Perry). These are boiled together in about a half-gallon of water until it turns dark brown; about 10–15 min. This will make enough to drink three times for 1 day, and should be consumed for 1 or 2 days.

For hemorrhages and heavy menstruation, one healer uses *q'eq xeeb'* (unidentified) and *kaq uuqub'* (unidentified; JD 81) leaves mashed together in water and boiled. The woman would drink this five times in 1 day. The woman would be massaged on every part of her body and the hammock would be used to "put the womb in place." For heavy periods, one woman says to boil the stem and leaves of *cola de alacran* (scorpion tail: thought to be *Heliotropium angiospermum* Murray) in three cups of water and drink one cup warm daily until menstruation is finished. A shifted womb can cause excess menses which is treated with massage and binding of the pelvic region, according to one midwife. Women are also not allowed to do heavy lifting during this treatment. Then five open *hibiscus* (*Hibiscus rosa-sinensis* L. var. *rosa-sinensis*) flowers and four closed *hibiscus* flowers are boiled and drank three times a day during the woman's period.

Don Francisco uses *xoy pim* (*Miconia oinochrophylla* Donn. Sm.; JD 78) to treat women who menstruate too frequently. The abaxial leaf surfaces are purple to fuchsia colored. Twelve leaves are crushed fresh in 1 L of cold water. This plant is considered a cold remedy in Don Francisco's classification system. If the plant is dried, the leaves could be boiled. The woman would drink this once in the morning and once in the evening, for 2–3 days. *The Handbook of Kekchi Medicinal Plants of Belize* says that *Miconia oinochrophylla* leaves are crushed in cold or hot water and the liquid is drunk to treat a heavy menstrual cycle (Maquin et al. 2005).

Amenorrhea (Absence of Menstruation: Primary and Secondary)

Amenorrhea is the lack of menstruation in women who are not pregnant or menopausal and should be menstruating. Primary amenorrhea is the delay of menarche or first menstruation in young women and secondary amenorrhea is the ceasing of a woman's menstruation, who is in her reproductive years, but not pregnant. In Q'eqchi' traditional medicine, when women are not menstruating it is because there is excess "cold" inside the girl or woman due to practices such as drinking too much cold water, or soaking in water during her period if she has already menstruated.

Don Francisco uses *chu pim* (*Hyptis verticillata* Jacq.; JD 12) to treat primary amenorrhea. The plant's name means "stinky herb" which it has been given due to its strong unpleasant scent. The plants are found in secondary forests, in house yards, or in fields like *wamil* (abandoned *milpa*) fields. The leaves are boiled and given to the patient warm to drink in order to "warm the blood, so that it flows." A handful of the leaves are boiled in two quarts of water. The woman would drink as a tea only in the morning for 3 days. She would drink once a month, for 3 days, until she starts her period. It depends on the individual how long this treatment takes to be effective. Some girls will start their period after the first couple of days after the treatment, and with other girls it takes some time. This plant could be stored for 4–5 months, so his patients could make the tea themselves, or come back to him monthly for him to prepare it.

Don Manuel uses *pereen pim* (*Psychotria poeppigiana* Müll. Arg.; JD 24), to treat primary and secondary amenorrhea. It is called *pereen pim* in Q'eqchi' because the bright red-orange flower bracts resemble a rooster's comb, and the word for rooster's comb is *pereen* and *pim* means herb or medicinal plant. In Kriol, the plant is called "hot lips" since the bracts are also "lip-like." The plant is common and grows anywhere he says. We collected the plant in secondary forests in a few different areas. Don Manuel uses the leaves to treat women when they are not getting their period, both when women have not reached menarche, and when women have their periods stop, but are not pregnant or going through menopause. This plant is considered a hot plant, according to Don Manuel's system of classification. The plant works he says by "opening the flow for the period." He collects leaves and boils them for the patient to drink as a tea, three times a day, until the woman begins menstruating. The plant is also used to treat hot flashes to "flow the problem out," he says.

The root and leaf of *yut it* (*Piper peltatum* L.; JD 15) along with *rix choql* (unidentified), *saq q'ehen* (unidentified; JD 48), and *saq kampaan* (unidentified) are boiled together by another healer to treat amenorrhea. This is taken as a hot tea for 3 days, and usually the period will come in less than a month he says. For secondary amenorrhea, he would use *rix choql*, *saq q'ehen* and *saq kampaan* as well "to soften hard blood" because "menstruation becomes hard because of "fright" (a culture-bound disease also called *susto* in Spanish and *kaanil* in Q'eqchi')."

For women experiencing primary and secondary amenorrhea, one midwife says that women should not drink cold drinks or bathe in cold water. She would treat this with the *quarum* (unidentified) tree which is boiled and drunk three times a day. For secondary amenorrhea, one herbalist uses a plant called *k'a che'* (unidentified; thought to be *Acosmium panamense* (Benth.) Yakovlev). Three pieces are boiled to bring on menstruation and the woman should drink as much as she wants.

Delayed Menstruation

Ms. Francisca and Ms. Adelina both agree late menstruation can be treated with jackass bitters (*Neurolaena lobata* (L.) R. Br. ex Cass.). For delayed menstruation, *serosi* (*Momordica charantia* L.; JD 69) is also used to "bring down a period."

One remedy uses *garlic* (*Allium sativum* L.), *serosi,* and *jackass bitters* (only three leaves of this plant) boiled together. The more you boil, the more bitter it will become and the stronger the medicine will be. *Momordica charantia* L. is also used to treat amenorrhea among Q'eqchi' women in Livingston, Guatemala (Michel 2006), as well as to treat "suspended menstruation" commonly among women in the Dominican Republic (Ososki 2004).

Another remedy uses one entire *b'enk* (basil: *Ocimum campechianum* Mill.) plant boiled in approximately four cups of water for 5–10 min. This tea can be consumed three times a day. For menstruation that does not come, "to bring down a period" a remedy made from *b'enk* root mashed and boiled with mashed *garlic* (*Allium sativum* L.) and *pom pim* (*Piper* sp.) leaves, for 5–10 min is used. The woman would drink three times a day until her period comes. These plants are also used for painful menstruation. A variation uses a bunch of *pom pim* and one *b'enk* plant boiled for 20 min. One herbalist would boil a small amount of *ruda* (*Ruta* sp.) branches and have the woman drink two times daily before meals. For delayed menstruation (and for abortion), another herbalist uses young *papaya* (*Carica papaya* L.), mashed up and put in boiling water, then strained and drank three times a day until the period comes. Drinking the astringent juice of papaya was also documented to be used for inducing abortion in Bangladesh (Sarker 1981).

There is certainly a gray area between remedies to bring on delayed menstruation and remedies for terminating a pregnancy. This is due largely to the fact that women often do not know for certain the reason for not menstruating at their expected time. In Cali, Colombia (Browner 1985b) reported that women considered regularity of their menstrual cycles to be indicative of good reproductive health and that some women utilized home remedies to ensure this regularity. She explains how for most Latin American women, delayed menstruation is an ambiguous sign; it may mean that she is pregnant, or just a natural change, or a menstrual delay known as an *atrazo* that is believed to be potentially harmful because the retained menstrual blood can lead to serious illness. So, women typically treat this ambiguous delay immediately to maintain good health, thus allowing for some covert fertility regulation. Browner (1985b) further explains that, "since it is difficult to distinguish a late period from an early pregnancy, unwanted pregnancies may be surreptitiously terminated through use of privately administered remedies. Women find this an effective and culturally acceptable way to practice fertility control."

2.2.3 Fertility, Infertility, and Fertility Regulation

Causes of Infertility

According to Don Francisco, there are two main reasons for female infertility. The first being that the womb and ovaries are "not clean so they are not producing eggs." The second major cause of female infertility in his opinion is when the womb is in a bad position or in a wrong position, which is thought to be caused most often by a bad fall. He says infertility can also occasionally be due to emotional problems.

Fig. 2.5 *Kok'mox* (*Maranta arundinacea* L.) leaves and flowers

Another male traditional healer said that female infertility can be due to the ovaries being "dry" and therefore not producing eggs.

Female Infertility Treatments

Infertility treatments are believed to wash the unclean ovaries or womb according to the male traditional healers. As an alternative explanation, a midwife said that infertile women need to have the "cold" from inside, taken out. If a lady cannot have a baby, she is massaged several times and *pimienta gorda* (*Pimenta dioica* (L.) Merr.) is boiled with *cinnamon* (*Cinnamomum* sp.) stick and given to the woman to drink to "take out cold from inside," according to this midwife. Among Qiché Maya midwives in Guatemala, there is a belief that "sterility may be caused by a 'cold womb' which consequently does not receive semen. One treatment is to warm the womb in the sweat bath and administer 'hot' herbal teas. However, if the sterility is caused by God, the midwife cannot cure it" (Cosminsky 1994).

If a woman's ovaries are unclean, Don Francisco uses *kok'mox* (*Maranta arundinacea* L.; JD 6; Fig. 2.5) to wash ovaries in treating women's infertility. The meaning of the name is "small *waha* leaf" since the plant looks like the *waha* leaf, which is used to wrap food. *Mox* means *waha* leaf and *kok'* means small. He uses at least 20 "heads" (bulbous swellings) of the roots, which he boils in a gallon of water. The woman would drink three cups in the morning and three in the evening, and three cups the following morning, so that she drinks three cups each at three different times. The gallon would be enough to complete the treatment. On the last day of her treatment, Don Francisco would have the woman lie in a hammock, with her feet pulled high up in the hammock, and massage her stomach. This plant is considered neither hot nor cold in Don Francisco's classification system. *Kok'mox* roots could be stored for a while because they do not rot easily he says. The plant is common, but grows only in primary forest, according to Don Francisco.

Next in his infertility treatment, he would have the woman drink *ixq pim* (*Tococa guianensis* Aubl.; JD 77). *Kok'mox* (*Maranta arundinacea* L.; JD 6) is only used to "wash out the ovaries," and the *ixq pim* and *tz'ulub'pim* (unidentified; JD 92) would be used for the 3 months after to "build up the eggs." *Ixq pim* is used by Don Francisco to treat female infertility and "clean the entrance for sperm" and used to "keep the sperm inside." The Doctrine of Signatures refers to human selection of the medicinal plants based on some physical characteristic of the plant that resembles a human body part or human body characteristic, with the belief that these plant characteristics are suggestive of the plant's medicinal application. *Tococa guianensis* Aubl. has ant-housing structures, known as domatia, at the apices of its petioles, which are thought to resemble human female genitalia, hence the name *ixq pim*, which means "female" herb. For female infertility, he boils 13 leaves with the domatia at the base of the leaves in 1 L of water for 10–15 min, which makes three cups for her to drink. He says that the 13 leaves correspond to 13 energy points in the body, one on the center of the forehead, and six on each side of the body at the sites of the wrists, elbows, shoulders, hips, knees, and ankles. After washing the ovaries with *ixq pim*, Don Francisco then does a 3-month treatment, each month starting when the woman's menstruation ends, using another very small plant of *tz'ulub'pim* (unidentified). He also uses *ixq pim* (*Tococa guianensis* Aubl.; JD 77) as a permanent male birth control, but he says the man should drink one cup, but it causes permanent sterility which cannot be reversed. This is a cold remedy when treating women and a hot remedy when treating men according to Don Francisco's system of classification. *Ixq pim* is pretty commonly found he says.

For female infertility due to emotional problems, Don Francisco would use 13 whole *tz'ulub'pim* (unidentified) plants with 20 root heads from *kok'mox* (*Maranta arundinacea* L.; JD 6) boiled and given to the woman to drink. This would go along with a massage. Don Manuel uses *kok'mox* as well to "wash the womb so that a woman can get pregnant." He uses the roots, which he says need to have "heads" (bulbous swellings) on them. He uses six or seven root "heads" and boils them for the woman to drink. She drinks one cup three times a day for 3 days. This plant is difficult to find, and grows mostly in primary forests, he says.

Don Manuel says that treating infertility is a process. According to him, either the woman is the problem, or the man, though it is usually unclear, so both should be treated. If Don Manuel is certain that only the woman needs to be treated, he would have her drink a tea of the boiled leaves of *kok'mox* (*Maranta arundinacea* L.; JD 6) or *kakaw che'* (thought to be *Alseis yucatanensis* Standl.) to "clean" or "wash" the ovaries, during the first quarter of the moon, the first month. Then every month during the first quarter, she would drink cold infusion or boiled tea of the leaves of *ixq pim* (*Tococa guianensis* Aubl.; JD 77). If the man is also being treated he will, starting this second month, also drink cold infusion or boiled tea made from the leaves of *teelom pim* ("male herb": *Clidemia crenulata* Gleason; JD 94). One herbalist knew that *ixq pim* was used for female infertility, but she was not sure of the preparation. For contraception, *teelom pim* was used, she knew, but again, she was not sure of the preparation.

To treat infertility, Ms. Francisca says *chalche* (unidentified) leaf, *manzanilla* (thought to be *Matricaria recutita* L.) stems, *che'oreek* (*Lippia graveolens* H.B.K.; JD 19) leaf, and *anise* (*Pimpinella anisum* L.) seed should be boiled together. One bottle is consumed every day. This is good for "cold inside" (the womb). If a woman wants to get pregnant, Ms. Adelina says she should cut up "one dollar amount" sold at the *cobanero* shop (approximately two tablespoons) of each of: *manzanilla* stems with leaves and flowers, *oregano* (thought to be *Lippia graveolens* H.B.K.) leaves, *chalche* (unidentified) leaves, along with *garlic* (*Allium sativum* L.) and three *cloves* (*Syzygium aromaticum* (L.) Merrill & Perry). These are boiled together in about a half-gallon of water until it turns dark brown, which takes 10–15 min. This will make enough for one day, and the woman should drink this three times a day, every day for 1 week that she does not have her period.

For infertility in women, one midwife recommends *cocolmeca* (*Dioscorea* sp.) soaked in *anisado*, which is a liquor made from *anise* (*Pimpinella anisum* L.) for 3 weeks and three tablespoons consumed daily. One herbalist said that *q'eq xeeb'* (unidentified) can be used to treat infertility. The plant is mashed, strained, and drank. "The woman would take a massage first and be put upside down in the hammock," (with her stomach facing upwards, but her feet high up in one side of the hammock so that her body is on a downward slope), she explained. After her period, she would drink *q'eq xeeb'* two times, once in the morning and once in the evening.

One traditional healer uses massage first, then *kun che'* (unidentified) roots are boiled for the woman to drink for 3 days after her period, for a woman who is infertile due to a "womb which is sideways." He says the same plant can be used for miscarriage prevention.

Female Contraception

Don Francisco knows five plants that work individually as contraceptives for women. Only one would be used and would be prepared and consumed every month after the woman's period. These plants are *xch'up ixim* (*k'otz'* or wild yam: *Dioscorea bartlettii* C.V. Morton; JD 18) stem and roots, *teelom pim* (*Clidemia crenulata* Gleason; JD 94), *aj tzo'pim* (unidentified), *xnaq' o* (unidentified), and *xnaq' paql* (unidentified). *Xch'up ixim* is a commonly known and used contraceptive in the Q'eqchi' community and was mentioned as a contraceptive by multiple traditional healers and herbalists.

Don Francisco recommends *xch'up ixim* (*Dioscorea bartlettii* C.V. Morton; JD 18) as a female contraceptive. The yam is "in the shape of corn" and *ixim* means corn. The name could be translated as "heart of corn," he says. The yam is cut up into pieces and boiled and drank as a tea. A piece about $1 \times 1 \times 4$ in. is cut into four pieces and boiled in a quart of water. After menstruation, the woman would drink this decoction three times; once in the morning, once in the evening, and then once again the following morning. This is done monthly. He says there are no other uses for the plant other than as contraception. This is considered a hot plant according to Don Francisco's system of classification. This plant is very hard to find, he says, and mostly grows in rocky areas in the hills in primary forests. Don Francisco has to travel far to find *xch'up ixim* not because of overuse, but because much of the forest has been destroyed.

This contraceptive is commonly used according to Don Francisco and some women do prepare it themselves. He says it is the most commonly used contraceptive among the Q'eqchi. According to Don Francisco, there is no danger in using this contraceptive for a long period of time. He says the woman will experience a "change in her scent," but there is no danger in using the plant. He says the contraceptive is effective if she drinks the tea every month. If she misses a month or two, she would get pregnant. The yam could be stored for a long time; one about the size of a softball may be enough to last about 6 months for use in contraception.

According to Don Francisco, *teelom pim* (*Clidemia crenulata* Gleason; JD 94; which means "male herb") is used as female birth control and to treat male impotence. Like *ixq pim*, *teelom pim* has ant-housing domatia, in this case at the base of the petiole that are thought to resemble human testicles, which is the inspiration for the plant's Q'eqchi' name. He says for treating both men and women, he would use 13 leaves and some of the "balls" (domatia). The 13 leaves correspond to 13 energy points in the body. It is considered a hot plant in his system of classification. The patient cannot have anything cold during the treatment. One liter of water is boiled and poured over the leaves and domatia and makes three cups of tea. He would have the woman drink one cup in the morning, one in the afternoon, and one in the evening, after her menstruation ends. If a woman drinks *teelom pim* for 1 day, the contraceptive effect would last 1 year to a few years, he says. She could reverse this effect by drinking three cups of *ixq pim* (*Tococa guianensis* Aubl.; JD 77). This is considered a hot remedy in Don Francisco's classification system. The plant grows in secondary forests in red soil. He says that it is less common than *ixq pim*.

For male impotence, the man would drink three cups of *teelom pim* (*Clidemia crenulata* Gleason; JD 94) tea, prepared in the same way as the woman's contraceptive. He would drink this during the quarter moon, one in the morning, one in the afternoon and one in the evening. The plant can be stored dry for 1 or 2 months and still be used. When asked why he thinks there is a clue on the plants (the human genital-resembling domatia) he said this is the way God made these plants, and he does not know why but his ancestors used these. He says these "balls" are the ants' home.

Teelom pim (*Clidemia crenulata* Gleason; JD 94) leaves are boiled and drank once a day, during the woman's period, for female contraception, according Don Manuel. The plant is considered a hot plant, because he says it is used to "kill off eggs in women." It is really a matter of reducing the chance of pregnancy he explained. Don Manuel warned that a woman may take this contraceptive plant for 2 years and then be permanently sterile, however for shorter periods she could take the plant, and then stop and still be fertile. *Teelom pim* grows mostly in red soil, sometimes in primary forest, but mostly in secondary forest, according to Don Manuel.

Another healer says *teelom pim* (*Clidemia crenulata* Gleason; JD 94) works as a female contraceptive because it "blocks the entrance to the woman's egg." He would use 13 "balls" (domatia) mashed and boiled in 1 gallon of water and given to the woman to drink for 1 day after menstruation ends. He says this would work for 1 year. He says 13 works, nothing less and nothing more will work. He only uses the "balls," and not the leaves. He says he uses the leaves of a plant called *puchuch q'ehen* (unidentified; JD 85) to reverse the effects of *teelom pim* by "opening the entrance to the egg for sperm." The woman would drink one cup three times with three

Fig. 2.6 Ms. Francisca slicing avocado seed to demonstrate contraceptive preparation

massages. He says there are two different plants with the name *puchuch q'ehen* and he would use the dark and light leaved ones for this treatment.

For birth control, Ms. Adelina would use about four big handfuls of *sorosi* (*serosi*: *Momordica charantia* L.; JD 69), and two leaves of *g'an mank* (jackass bitters: *Neurolaena lobata* (L.) R. Br. ex Cass.) boiled in a half-gallon of water for 10–15 min. This would be consumed 1 day during the woman's period or throughout the duration of her period (depending on plant supply) three times a day. Ms. Francisca says that *sorosi* tea can be consumed three times a day before a woman's period and throughout the period. For birth control, Ms. Francisca said women could boil chopped avocado (*Persea americana* Mill.) seed in four or five cups of water for 15 min, and drink during menstruation (Fig. 2.6). The water should turn bright red once the seed has been boiled long enough. Younger women in Ms. Francisca's family use *avocado* seed as a contraceptive. According to one herbalist *avocado* seed is more effective than *cowfoot* (*Piper auritum* H.B.K.; JD 27) because *cowfoot* can cause permanent sterility, which she says will not happen with *avocado* seed. She says she uses *cowfoot* though because you cannot always find *avocado* seed, since the season in Belize is only a few months long. *Persea amercicana* Mill. seeds are also used as contraception and for sterilization in women in northern Peru (Bussmann and Glenn 2010), as a contraceptive in Oaxaca, Mexico (Browner 1985a) and in western Guatemala (Cosminsky 2001), and a tea made from the seed was used to bring on menstruation in Cali, Colombia (Browner 1985b). For contraception, one herbalist would boil a large handful of *avocado* leaves ("pear" leaves as they say in Belize) until the water is very brown and drink three times on the last day of her menstrual period.

Ob'el (*pata de vaca*, cowfoot: *Piper auritum* H.B.K.; JD 27) can be used as birth control, though some believe it could cause permanent sterility. One handful of leaves is boiled and drank before meals during the woman's period. For contraception, one herbalist uses the inflorescences of *ob'el*. She cuts up, mashes, and extracts the juice of 6–8 inflorescences. She boils them for 5–6 min. She would drink this once every morning before meals during her period, or for 1 day, 2–3 times that day.

Tuni cos (unidentified) roots can be used as a contraceptive, according to one traditional healer. The roots are cut up, toasted, and then boiled and the decoction consumed after the woman's menstruation. Just a little should be consumed just once. *K'aham* (unidentified, however Ms. Adelina says is the same as *cha'jom k'aham*, which is *Gouania lupuloides* (L.) Urb.) can be used as birth control. *Winq pim* (unidentified) is another she said, but it is found deep in "the bush" (forest) and she does not know much about it. For birth control, one herbalist says *tzo'pim* (unidentified) can be used. Three 6-inch pieces of the stem with leaves should be boiled and drank once in the morning, once in the afternoon, and once in the evening.

Pregnancy Test

The leaves of *frijolillo* (thought to be *Senna occidentalis* (L.) Link) can be used as a pregnancy test, by urinating on the leaves. If the leaves "look scorched" then it is thought to be positive indicating the woman is pregnant, and if the leaves look normal then it is negative.

Miscarriage Treatment

Ms. Adelina would cut up "dollar amounts" (approximately two tablespoons) each of: *manzanilla* (thought to be *Matricaria recutita* L.) stems with leaves and flowers, *oregano* (thought to be *Lippia graveolens* H.B.K.) leaves, *chalche* (unidentified) leaves, along with *garlic* (*Allium sativum* L.) and three *cloves* (*Syzygium aromaticum* (L.) Merrill & Perry) to treat a miscarriage, (or postpartum hemorrhage or heavy periods). These ingredients are then boiled together in a half-gallon of water for 10–15 min, until it turns dark brown. This will make enough to drink three times for 1 day, and the woman would drink the remedy for 1 or 2 days. Another amount would be made for the second day if needed.

Abortifacients

Not everyone was asked about abortifacient plants during the interviews. Everyone was asked about contraception; however, if the person seemed uncomfortable being asked about contraceptive plants, or seemed as though they might be uncomfortable or offended by the topic in any way, they were not questioned about abortifacient plants. Some people with which I have developed a rapport, and I felt would be comfortable with the topic, have been asked specifically. Others offered up the information during the course of their interviews, without the topic being mentioned to them. In general, one healer explained that Q'eqchi' people do not feel good about abortion, so often it is done in secret. The husband would give consent usually, and would go with his wife to seek the treatment. In some cases, people secretly terminate pregnancies from adulterous relationships as well.

For abortion (or delayed menstruation), *garlic* (*Allium sativum* L.), *sorosi* (*serosi, yamor*: *Momordica charantia* L.; JD 69), and three leaves of *jackass bitters* (*g'an mank*: *Neurolaena lobata* (L.) R. Br. ex Cass.) are boiled together, until the water is dark. The more it is boiled, the more bitter it will become and the stronger it will be. For abortion up until the first month of pregnancy, one cup is consumed three times a day, and is said to usually work within 3 days. If the pregnancy is farther along, 2–3 months pregnant, then this would be consumed all during the day. This is a widely known and perhaps commonly used recipe for abortion. Another variation uses a bunch of *sorosi* (2–3 handfuls), and 4–5 leaves of the "wide kind" of *jackass bitters* (variety with wide leaves), or 12 leaves of the "narrow kind" of *jackass bitters* (variety with narrow leaves). These would be boiled in a gallon of water with the *sorosi* until the water gets dark. This entire gallon would be consumed throughout the day until it was finished. Other variations of this remedy are common. One is made from just a large bunch of *sorosi* boiled and the liquid consumed. Another variation uses just *jackass bitters*, in which a bunch of leaves are boiled and the liquid consumed 4–5 times during the day. This is said to be effective by the same evening. Another boils four leaves of *jackass bitters* and a whole *b'enk* (basil: *Ocimum campechianum* Mill.) plant. Just a cup and a half is consumed. Only one abortifacient recipe did not contain *Neurolaena lobata* (L.) R. Br. ex Cass. or *Momordica charantia* L., which was young *papaya* (*Carica papaya* L.) mashed up and put in boiling water, then strained and drank three times a day until the period comes. This is also for delayed menstruation.

2.2.4 Conception and Pregnancy

Don Francisco noted that he can tell if a woman is pregnant because the areas right above her eyeballs will sink in. When a woman is pregnant with a girl, the woman gets pale, and when a woman is pregnant with a boy, she gets darker. One midwife says she can tell when someone is pregnant by looking at them because there are facial signs, blood changes the color of the face, and the facial features change. When the woman is pregnant with a female baby, she becomes pale with sunken eyes. When she is pregnant with a male baby, her eyes and face are more pink or red. She tells her patients the sex of the baby during pregnancy. Another midwife said she can tell women are pregnant before they are showing. "The woman will look paler, her hips are more open from the back, and some women like to spit a lot, or sleep too much or eat too much," she says. Sometimes she can tell if someone is going to have a boy or a girl. If the woman is going to have a girl, her breathing is slower, and if the woman is going to have a boy, her breathing is faster. If the mom is having twins, this midwife can feel two heart beats.

Beliefs Regarding Pregnancy

There is a belief that pregnant women should not stay in the sun too much because this will dry the placenta. Pregnant women should not go by the fire hearth for baking (regular cooking is ok) or go out in the bright sun because the placenta will dry and not come out easily. She should wash clothes in the shade not in the bright sun. If a pregnant woman leaves dish water in the basin and does not toss it out, the baby will be in too much "water" (amniotic fluid) and her pregnant stomach will get very big. During pregnancy women should not bathe too late in the day and should avoid cold foods according to Ms. Francisca. One woman said she believed that pregnant women should not eat *jippi jappa* (*Carludovica palmata* Ruiz & Pav.) or *chaya* (*Cnidoscolus chayamansa* McVaugh) because the child would be born too weak. She said they also cannot eat yesterday's food. When a pregnant woman is eating, if she tries to scrape up the last bits from the pan to eat, her baby will have a large head. If a child asks a pregnant woman for food, she must give it to the child or her baby will come out hand first. When a woman is pregnant, if she reaches and stretches too much this will make the umbilical cord too long and wrap around the baby's neck. If a pregnant woman puts a piece of wood with a "V" shape in the fire and then turns it around, "the baby will come breech."

Awas refers to a variety of illnesses or characteristics a child could acquire in the womb if the mother breaks a social rule; the word can be literally translated as "bad luck," "sorcery," or "taboo" (Kahn 2006). A pregnant woman should not "scorn" (laugh about or say negative things about) people with retardation or disabilities. She should not make fun of someone, because her child could have the same problem that she made fun of. If there is a food that a pregnant woman scorns or craves, her baby could have an *awas*, in this case a birthmark that would come out looking like the food that was "scorned." If a baby is born with *awas*, a piece of the person's clothes or hair that was "scorned" would need to be burned under the baby after birth. If the person or thing that was "scorned" cannot be found, a type of plant, a "prickle" (unidentified) is burned instead.

One herbalist said that each of her children had been born with an *awas* and she had to think back to what she might have "scorned" or done to cause these, to find the source for the remedy. Her one son was born with two differently colored testicles, and this woman remembered having laughed at a pig with two different colored testicles. She found that same pig and burned some of its hair under her baby and the coloring of her baby's skin evened out. Another one of her sons could not urinate when he was born. She had collected water every day while pregnant from a container that she would plug by putting a small stick in the hole the water drained from. When her son was born with the *awas* and unable to urinate, she burned that same stick under the baby and he was able to urinate properly again.

Another herbalist says she treats things such as "when a woman is pregnant and sees a dying cat for example, and her baby would act like this" (like a cat after it was born, presumably crying and fussy). If a pregnant woman sees a newborn baby and does not hold the baby, the baby will stool green, and have heavy breathing, because the spirit of the baby in the pregnant woman's womb is "heavy." If this happens, the pregnant woman will have to come hold the baby or make the sign of the cross on the baby's forehead with her saliva. Women who are pregnant should avoid staying out too late and "catching spirit." This is caused by the "winds" at midnight, also at midday and in certain parts of the forest. To treat *awas*, when someone "catches spirit" due to the wind, two live fish are caught and passed over the woman and the fish are sacrificed.

Miscarriage Prevention

Don Francisco uses prayer and massage for preventing miscarriage, and Don Manuel says he uses prayer to put the fetus back in place. Women who fall and experience pains during pregnancy are treated with prayer and massage. One healer said he would start from the fourth month of pregnancy, positioning the womb to avoid miscarriage if the woman is having pains. To prevent a miscarriage when a woman has fallen and the baby has shifted, massage and banding of the woman's abdomen would be used according to one midwife.

Don Francisco uses *san saar q'ehen* (*Aristolochia tonduzii* O.C. Schmidt; JD 2) to treat a woman who starts experiencing bleeding during pregnancy, e.g., during the fifth or sixth week. The plant's name means "incense burner" and is called such because the dried fruit has the shape of an incense burner. To treat "pregnant menstruation," 12 or 20 leaves can be boiled to make a tea. Or, using the stem as well, maybe a foot of stem is cut up into pieces along with four or five leaves to make the tea. The woman drinks one cup in the morning and one cup in the evening and then the treatment is complete. The plant is also used to treat postpartum pain. The plant is considered a hot plant according to Don Francisco's classification system. This is a common plant that grows mostly in very good soil on hillsides in the primary forest, although a few plants can sometimes be found in secondary forests. The plant could be dried and stored to be used for over a year, he says.

Another species Don Francisco uses to prevent miscarriage is *xoy pim* (*Miconia oinochrophylla* Donn. Sm.; JD 78). The abaxial leaf surfaces are purple to fuchsia colored. This remedy is recommended for women "who always miscarry." Twelve leaves are crushed fresh in 1 L of cold water. The woman would drink this once in the morning and once in the evening, for 2–3 days. If the plant is dried, the leaves should be boiled. This plant is considered a cold remedy in Don Francisco's classification system.

To prevent miscarriage, *cocolmeca* (*Dioscorea* sp.) *root* should be chopped up and mixed with *ginger* (*Zingiber officinale* Roscoe.) and boiled together and drank three times daily. Miscarriages and preterm births can be caused by infections. Both *Dioscorea* and *Zingiber officinale* have been shown to be estrogenic (Cheng et al. 2007; Kang et al. 2006) as well as antimicrobial (Kelmanson et al. 2000; Joe et al. 2009). For a woman who bleeds during pregnancy, she can drink *cocolmeca* boiled with *ginger* or *fever grass* (*Cymbopogon citratus* (DC.) Stapf.) "to prevent the baby from coming out." The pregnant woman would then get massaged three times using strokes with an upward motion. The *Chinese Herbal Medicine: Materia Medica*

(1993) states that *Zingiber officinale* Rosc. "warms the channels and stops bleeding: for cold from deficiency that may present with hemorrhage of various types, especially uterine bleeding. It is used in treating hemorrhage *only* if the bleeding is chronic and pale in color, and is accompanied by cold limbs, ashen white face, and a soggy, thin pulse," though states to "use with caution during pregnancy."

One midwife had recently treated the threat of miscarriage in a woman who did not know she was pregnant and drank *sorosi* (*serosi*: *Momordica charantia* L.; JD 69; this plant is used as an abortifacient and for maintaining general health). The midwife gave her a *hibiscus* (*Hibiscus rosa-sinensis* L. var. *rosa-sinensis*) leaf remedy and massage. If miscarriage occurs, the woman should drink *yerba de cancer* (*Acalypha arvensis* Poepp.; JD 54) 2–3 weeks after, to take out whatever infection is inside.

Pregnancy Discomforts

Don Francisco uses *b'aknel pim* (*Drymonia serrulata* (Jacq.) Mart.; JD 16) to treat pregnant women who experience pain with very mild fever. According to him, *b'aknel* is the name of snake similar to the "yellow jaw" (*Bothrops asper*) but is the smaller version, and *pim* means "herb." The flower of this plant is in the shape of the head of that snake, he says. It is an epiphyte that can be found on decaying logs or in very moist areas on rocks. Thirteen or twenty leaves would be crushed in 1 gallon of cold water and then would have warm water added to it. This is done to keep the warmth inside the pregnant woman. She would drink some and bathe in the rest of the water. This plant is considered a cold remedy according to Don Francisco's system of classification. The plant must be used fresh since it decays easily.

To treat back pain associated with pregnancy or after pregnancy, Don Manuel uses *yut it puchuch* (thought to be *Piper* sp.). The leaves are boiled for about 30 min and drank as a tea, one cup in the morning and one cup in the evening, until the person gets better. This plant is used with *markos q'ehen* (*Piper pseudolindenii* C. DC.; JD 25), but could be used individually, and is prepared in the same way. This is considered a hot plant, in Don Manuel's classification system. He says the plant is difficult to find and grows in primary forests in hilly, cold areas. One herbalist says that a tea from *provision bark* (*sapaton* in Spanish: thought to be *Pachira aquatica* Aubl.) should be drank during pregnancy to "have enough blood."

Morning Sickness

For morning sickness, Ms. Francisca recommends drinking water with mashed *garlic* (*Allium sativum* L.) in it, in the morning. Or drink a stout every few days. *Tamarindo* (*Tamarindus indica* L.) fruit can be chewed with a bunch of salt and pepper to alleviate morning sickness. One woman says for nausea during pregnancy and morning sickness she would boil *orange* (*Citrus sinensis* (L.) Osbeck) or *lemon* (*Citrus limon* (L.) Burm. f.) leaf, using ten leaves to make two or three cups.

Breech Pregnancy

According to one midwife, breech pregnancies occur when no one is attending to the pregnant woman, with prenatal massage and other care. In the last 2 months if the baby is not in position, she would attend to the woman frequently. If she started to treat in the fifth month of pregnancy, she would check on her every 15 days. One male healer says that he can turn breech babies late in pregnancy at 8 or 9 months using prayer and massage. To turn breech babies, another male healer says he would use "bush medicine" and prayers, but said his midwife wife does it differently. Her job is to put the child in the correct position. This starts at 3 months of pregnancy and continues throughout the pregnancy.

2.2.5 Birth and the Postpartum Period

Q'eqchi' men often deliver their wives' babies, especially in the more remote villages. This was not traditionally done in Q'eqchi' culture, but is now done by necessity in Belize because of the lack of traditional midwives in the villages. The husband's or wife's parents or grandparents sometimes instruct the husband on what to do or what he will need to do when the time comes. Four women from this study gave birth at home with the help of their husbands (two included the help of her mother-in-law as well). Three women gave birth at home with the help of their mother. Two older women had given birth at home with a midwife's help and one gave birth at home with the help of traditional healers. One of the male traditional healers had delivered his own children. One woman had both babies in the hospital because they were by Cesarean section. Younger Q'eqchi' women tend to go to the Punta Gorda hospital to give birth if they live in a village close enough to town. Though they frequently admit that they are unhappy with the quality of care they receive there.

Facilitating Labor

Don Francisco uses *xch'up al* (*Cissampelos tropaeolifolia* DC.; JD 17) to induce labor or to speed up a prolonged labor. The plant is named after the umbilical cord and bellybutton. The peltate leaf is shaped like a belly button, attached to a "cord like" petiole and vine. *Xch'up* means bellybutton, as well as umbilical cord. *Al* means child, so the name means "child's umbilical cord" or "child's belly button." The plant grows mostly in fields or primary forest of one region of Toledo especially, according to Don Francisco. The plant helps a woman to give birth when she has a birth delay because her system does not have sufficient liquid, he says. So this plant is used to build that liquid to quicken delivery. Nine leaves are crushed in cold water because the woman carries the baby for 9 months. The first cup is given and if it works, then the treatment would not need to be repeated. If it does not work, then the drink is given three times. After she takes the medicine, maybe 10 min after, Don Francisco would massage the woman to "fix the baby in the position for her to

Fig. 2.7 Ms. Francisca grinding hibiscus leaves with water on a *ka'* (grinding stone)

deliver." This is considered a "cold" plant according to his system of classification. The plant should only be used fresh because it is the slippery content that is needed for the treatment. When you crush the leaves it is very slippery. According to Don Francisco, this plant is getting more and more difficult to find because whenever the plant is burned or cut down it takes a long time to grow back again.

Cissampelos tropaeolifolia DC. is also used by the Q'eqchi' Maya in Livingston, Guatemala to release the placenta after childbirth. A handful of fresh leaves is infused in warm water and given to the woman as a tea in three consecutive doses (Michel 2006). In the Petén region of Guatemala, *Cissampelos pareira* L. leaves are boiled and the water is used to bathe newborn babies to treat painful navel (Comerford 1996).

Ms. Francisca says *hibiscus* (*Hibiscus rosa-sinensis* L. var. *rosa-sinensis*) leaves can be mashed and the slimy juice that results can be drank to speed labor (Fig. 2.7). She says a tea of *cloves* (*Syzygium aromaticum* (L.) Merrill & Perry), *cinnamon* (*Cinnamomum* sp.) stick, and honey can also be given to the woman to drink to speed labor. *Lime* (*Citrus aurantifolia* (Christm.) Swingle) root, *pericon* (thought to be *Tagetes lucida* Cav.), *romero* (*Rosmarinus officinalis* L.), *manzanilla* (thought to be *Matricaria recutita* L.), and *lime* rind (from a lime that is past its due date) are boiled together for about 15 min. This should be consumed like water all the time. Her daughter chimed in that she made it for her daughter-in-law who drank it in the morning and her labor pain started in the afternoon. After that they gave her *hibiscus* to drink.

According to Don Manuel, *wild hibiscus* (*Malvaviscus arboreus* Cav.; JD 26) flowers are crushed in water and given to the mother to speed delivery. A handful of the leaves are boiled and drank warm to treat a delayed birth. The *slipperiness* of the plant is used. It is difficult to find this plant since it grows at the base of mountains, so common *hibiscus* (*Hibiscus rosa-sinensis* L. var. *rosa-sinensis*) leaves could be used. Another traditional healer uses *ru'uj b'ach* (unidentified; JD 49) to speed labor. The leaves are mashed in water and given to the woman to drink and then her abdomen is massaged. "After 10 min the baby comes down." *Ru'uj b'ach* (unidentified) is a plant that is slippery and used to speed labor according to one midwife. When a baby is overdue, a handful of *purple maguey* (unidentified, thought to be *Tradescantia spathacea* Sw.) leaves should be boiled and the woman should drink as often as necessary until the pain starts, according to one herbalist.

Don Manuel says the leaves of *aq'l* (*po'jor*: *Cecropia* sp.) can be boiled and drank as a tea to quicken delivery. For difficult labor, Ms. Francisca uses young *po'jor* (*Cecropia* sp.) leaves, *tuulux* (unidentified) bark, and *q'eq xeeb'* (unidentified) leaves boiled together. This is for when the cord is around the baby's neck. Another treatment calls for boiling just one leaf. A tea of three young *Cecropia peltata* L. leaves is also used by the Q'eqchi' of Livingston, Guatemala to increase labor contractions during childbirth and to expel the placenta (Michel 2006).

Rismalhal ("hair of corn," corn silk: stigmas of *Zea mays* L. subsp. *mays*) from four ears of corn is boiled and one cup is given to the birthing woman to speed contractions. To ease delivery, one cup of the juice made from a fresh pad of *scoggineal* (*Opuntia cochenillifera* (L.) Mill.) is given to the woman to drink. Another treatment uses a small amount of *ruda* (*Ruta* sp.) branches boiled to ease childbirth and aid contractions.

One healer uses *mesb'eel* (*Sida rhombifolia* L.; JD 45) to speed labor. Two handfuls of the leaves are mashed up in water (not boiled), and one cup is given to the mother to drink and in 5–10 min the baby should come down, he says. *Sida rhombifolia* L. is used by the Q'eqchi' Maya of Livingston, Guatemala as a tea made from a warm infusion of the leaves to ease the pain of childbirth (Michel 2006). This particular healer also uses a plant called *b'ach* (unidentified) to speed labor. He would collect a bunch of the young branch tips with leaves from approximately 20 plants. He would mash the leaves in water and give the woman one cup to drink.

To speed up labor and help with labor pains, one herbalist uses honey and six *cloves* (*Syzygium aromaticum* (L.) Merrill & Perry) and *lime* (*Citrus aurantifolia* (Christm.) Swingle) rind, mixed with water and given to the woman to drink. Or she says to mix honey with *cinnamon* (*Cinnamomum* sp.) stick and drink the water, and this will aid in her labor. During labor, to reduce pain and speed the labor, ground *black pepper* (*Piper nigrum* L.) and *spice seed* (allspice: *Pimenta dioica* (L.) Merr.) can be given to the woman to drink with *cacao* (*Theobroma cacao* L.) or *tea* (*Camellia sinensis* (L.) Kuntze). For labor pains, she also suggests *sorosi* (*serosi*: *Momordica charantia* L.: JD 69). She says to boil the plant until the water turns yellowish or darkish, and drink it warm. One of the midwives explained that *lime* roots, three sticks of *cinnamon*, and nine *cloves* are boiled together in a small pot, and three cups are consumed daily mixed with honey. This brings labor pain if the woman's baby is overdue and she is not getting labor pains. Or *lime* rind grated and boiled with three sticks of *cinnamon* and nine *cloves* to "bring down pain" (bring on labor pain or in other words induce labor).

To increase contractions and speed labor, after the belly is massaged downward, a "dollar amount" (approximately two tablespoons) of *oregano* (thought to be

Lippia graveolens H.B.K.), a "dollar amount" of *manzanilla* (thought to be *Matricaria recutita* L.) and a teaspoon of *ab'aq* which is the "dark stuff that gathers on the thatch" inside a thatched house (thought to be soot). This can be rubbed from the beam above the cooking fire. If not, the burnt stuff under the *comal* could be used. These would be boiled together in about a half-gallon of water and drank. The same remedy would be prepared to expel a placenta, except that salt would be added. This remedy would have a massage that goes along with it. The dark colored "dust" under the *comal* is gathered and mixed in a little water and is massaged on the pregnant woman's abdomen so that the baby can come down quickly, according to one of the midwives.

Salt is placed in the woman's hand if her baby cannot come down and all other methods have been tried. It is believed that it happens because someone wished the pregnant woman evil so the salt helps to break that spell. Sometimes a baby is "bound" to a woman and she cannot deliver (does not go into labor). If this is due to witchcraft, one traditional healer said he would work at midnight and he would use plants to bring on her labor.

Beliefs Regarding Birth

According to one midwife, when a baby comes down with the foot first instead of the head the woman is put "upside down" in a hammock, with her stomach facing upwards and her feet far up one end of the hammock. The woman is lifted up and down several times (by the feet). After that, her abdomen or actually the baby inside her is massaged in a circular motion until the baby's head is downwards. She is also given a big cup of ground *black pepper* (*Piper nigrum* L.) in hot water to drink. When a baby's hand comes out first it is believed that when the mother was pregnant, she was very selfish. So a 5-cent coin is put in the baby's hand and it goes back in, and shortly after the baby is born.

Postpartum Practices

During the postpartum period, women should stay inside and be warmed for 3 days and 3 nights after giving birth, according to Ms. Francisca and many of the other Q'eqchi' women interviewed. A small area of the house should be set aside for the mother and the baby, perhaps by a partition or blanket. It is important to keep women's bodies warm after giving birth so they will not experience postpartum pains. During the first 3 days after birth, hot coals or hot stones should be kept under the mother's bed, and she should not go outside "where the breeze blows," or her health could decline. It is believed that if she does, the wind will enter her and cause her stomach to grow in size, causing her to have severe pain and cramps during menstruation or during menopause. According to Ms. Francisca, the mother should not go "into the winds" (outside) for one whole week because it could enter her ear and she may not be able to hear properly after. She says for 1 month the child should not be taken "into the wind or into the night." She says cotton should be put in the mom's ear and the baby's head should be kept covered. Others explained that the newborn baby should be warmed by being swaddled in a warmed cloth.

Shortly after birth, for 1 week, especially in the first 3 days, a rock that has been warmed in the fire hearth is wrapped in a cloth and used to warm the woman's belly, to ease postpartum pain, bring her uterus back to normal size, and help the uterus to heal. The rock will be rubbed and pressed all over the woman's abdomen. For pains after childbirth a stone should be warmed to massage the woman while she lies in a hammock. Her feet are pulled up high in one end of the hammock so she is lying with her head close to the middle of the hammock. In Don Manuel's words, "the womb needs to be warmed just after delivery to stop from bleeding too much. In the past they used medicinal plants. This needs to be done due to the enlargement of the womb so it needs to be heated. Now they use a 3–5 lb rock that is wrapped and put on the woman's stomach or side after the rock has been warmed. Her sides need to be warmed too. This is not really practiced by everyone anymore, so a lot of the women now have big hanging stomachs." There should be two rocks, one to be using, and one to be warming, as this is done pretty continuously throughout the first 3 days postpartum.

One of the midwives said she would usually massage a woman after birth, and band her abdomen with a cloth so the womb does not come out. The belly should be banded for 3 days after birth to avoid hemorrhage and bring down the size of the uterus. Banding or, alternatively, prayers should be used, according to another herbalist. Nine days after birth, the woman should be massaged from toes to head, and then have her abdomen banded, according to one herbalist. Another said the womb should be banded for the whole first week.

Ms. Francisca feels that women should eat only chicken, egg, beef, and other warm foods, no cold foods like cabbage or tomatoes for 1 month after giving birth. According to another woman, mothers should drink only chicken *caldo* (broth or soup) after giving birth. She says that everyone eats it after a birth; it's a tradition. For the first month after birth, some believe no pork should be eaten.

Postpartum Treatments

Don Francisco uses *san saar pim* (*Aristolochia tonduzii* O.C. Schmidt; JD 2) to help a woman who suffers a lot of pain after birth. It is used to heat the womb to relieve that problem, using a tea and a bath. To treat postpartum pain, the leaves and stem of the plant are boiled. For the tea, 12–20 leaves, or approximately a foot of the stem cut up and four or five leaves are used. For the bath, it does not matter; any amount of leaves could be used. The woman would bathe in the morning in a bath of the leaves and she would drink the tea, twice a day, morning and evening for 3 days. This plant could be dried and stored and used for over a year, according to Don Francisco.

Don Manuel would use *markos q'ehen* (*Piper pseudolindenii* C. DC.; JD 25) leaves boiled and drank warm to help take out the cold that entered during labor.

Fig. 2.8 *Ob'el* (*Piper auritum* H.B.K.) leaves and inflorescences

For postpartum paleness, if a woman has lost too much blood during childbirth, and postpartum swelling (which usually occurs a few days after) Don Manuel uses *ob'el* (cowfoot: *Piper auritum* H.B.K.; JD 27). Women get this because a lot of cold enters their bodies during birth he says. This plant is fairly common and is found mostly in secondary forests and in *wamil* areas (areas which were previously *milpa*, but have been taken over by successional vegetation). He uses the leaves of the plant, collects 20–25 of them, and boils them for 45 min to 1 h with the lid on the pot. The lid is not to be opened until the treatment because the steam is what helps the woman. He administers this remedy, by having the woman lay in a hammock with something covering her body, and the pot of boiled leaves placed underneath the hammock. The woman should have this treatment three times, once per day, and also cannot take a cold bath for 1 week. This plant is considered a hot plant, by Don Manuel's classification system.

A common practice among Q'eqchi' women is the use of *ob'el* (cowfoot: *Piper auritum* H.B.K.; JD 27; Fig. 2.8) leaves in a postpartum vaginal steam bath. One young woman who had her children in the hospital, said once she was at home she sat over a bucket of *ob'el* steam once a day for 3 days after the birth, in the evening when she would not be going out of the house again. *Ob'el* is used as a postpartum treatment ideally approximately 1 h after the birth of the baby and placenta. Three to five big leaves are boiled in a pot with about 1 gallon of water. This is then poured into a small bucket and the woman would sit on this until the water is barely warm anymore (about a half an hour). This is done once a day for the first month. One herbalist believed that if this is done early in the morning, the woman would be able to get pregnant easily again the next time. This brings out the "wind" and the cold and doing this for a month will "bring the belly back down," and "bring the womb back to place." According to one of the midwives, *ob'el* vaginal steam bath is also used to clean the womb postpartum and "take out all the clots and infection, and helps also to ease the pain after birth and get the womb back to normal."

Another herbalist says that after a woman gives birth, four or five leaves of *ob'el* (*Piper auritum* H.B.K.; JD 27) plus four or five small leaves of *b'enk* (*Ocimum campechianum* Mill.) are boiled in a small two-cup pot until the water is yellowish, which is then let to cool and consumed right after the baby is born. This prevents infection and pain, takes out the clots, the cold and the placenta. After childbirth, one woman says a large handful of *santa maria*, the "female" one she specified (thought to be *Piper auritum* H.B.K.) should be boiled. The woman then sits over the steam for 25 min. This helps put the uterus back in normal position and prevent infection and excessive bleeding. The "male" *santa maria* (thought to be *Piper peltatum* L.) is used as a tea after childbirth for healing from pain, swelling, and cold.

Ms. Francisca says a tea made from *manzanilla* (thought to be *Matricaria recutita* L.) mixed with *che'oreek* (oregano: *Lippia graveolens* H.B.K.; JD 19) should be drank for the first week postpartum by both mom and baby. One woman uses *kaxlan q'ehen* (*pimienta gorda*: *Pimenta dioica* (L.) Merr.) to heal the womb after childbirth (also for vaginal infections). Ten to fifteen leaves are boiled in a gallon of water for 5–10 min. The woman would sit on a bucket containing the liquid while it is hot, and sit there until it cools down. This would be done once every morning or twice for the first week or two after childbirth. This heals the womb and takes out the clots from inside. After birth, women should be given a drink of boiled *chalche* (unidentified), *manzanilla* (thought to be *Matricaria recutita* L.), and *garlic* (*Allium sativum* L.), according to one woman. A tea made from *samat* (*culantro*: *Eryngium foetidum* L.) roots boiled with *garlic* is drank after birth for pains.

Sorosi (*serosi*, *yamor*: *Momordica charantia* L.; JD 69) is used according to one herbalist for 1 month after birth for the final cleansing and to build the blood back up. One handful of *sorosi* is boiled with *garlic* and drank three times a day for 1 week. *Mesb'eel* (*Sida rhombifolia* L.; JD 45) is used to treat women after they give birth. Three whole plants approximately 1 ft high are cooked with three cups of water and three cups are consumed for the day. This is done for 3 days, starting the day of the birth. This will bring the womb back to size and take out any infection.

To warm the womb after delivery, one midwife uses three plants, a handful of *may pim* (unidentified), *che'oreek* (oregano: *Lippia graveolens* H.B.K.; JD 19), and four leaves of *yut it puchuch* (unidentified, though thought to be *Piper* sp.) boiled in a pot of water. The woman should drink a warm cup of the tea three times a day for 3 days. She says that postpartum massage is necessary because if not, the womb may go to one side and during the next pregnancy the baby may be malpositioned. One healer gives a postpartum massage to put the womb back, and ties it (bands the abdomen with a cloth), and has the woman drink a tea of and bathe with *puchuch q'ehen* (unidentified; JD 85) leaves. She should not put the water on her head. She should wait 3 days postpartum to do this and then do it for 2 days.

Ms. Francisca says for after the delivery of twins, three young leaves of *ox* (*coco*: *Colocasia esculenta* (L) Schott) are cut up with three green bird pepper (*Capsicum* sp.), and three small *corn* (*Zea mays* L. subsp. *mays*) tortillas are toasted (Fig. 2.9). The *ox* leaves are boiled with the *bird pepper* and cut up, salt added and eaten. After two babies come out she says this should quickly be prepared for the mom to eat. This treatment is for the pain following the birth of twins specifically. *Yerba de cancer* (*Acalypha arvensis* Poepp.; JD 54) should be drank after a C-section until the cut is cured to take out any infection, according to another herbalist. Spanish dry white cheese sold at the market (in Guatemala, maybe not in Belize) that is shaped like a rock but crumbles, should be eaten with toasted *corn* tortilla, according to one herbalist, to help heal a C-section.

Fig. 2.9 Bird pepper fruits and young inrolled *ox* leaf (with mature *ox* leaf in background)

Placentas

To expel the placenta, Ms. Francisca says that sometimes the end of the birthing woman's ponytail is put down her throat, or she is given salt to hold in her hand. If a woman has not delivered her placenta by 30 min after the birth, one midwife said she would gently knock the back and top of the woman's head, then push the woman's own ponytail down her throat, until she wants to gag and pushes out the placenta. The practice of putting the birthing mother's ponytail down her throat to help her expel the placenta is also found among the Qiché Maya in Guatemala (Cosminsky 1994). To help expel a placenta, another midwife would get an empty bottle and have a woman kneel down and blow into the bottle.

One healer uses *mesb'eel* (*Sida rhombifolia* L.; JD 45) for placenta expulsion. Two handfuls of the leaves are mashed up in water (not boiled), and one cup is given to the mother to drink. In 5–10 min the placenta should come down. This healer also uses a plant he calls *b'ach* (unidentified) to expel the placenta. He would collect a bunch of the young branch tips with leaves from approximately 20 plants. He would mash the leaves, not boil them and give the woman one cup to drink. The plant has red flowers, which can be seen in the dry season, he says.

If women give birth at the hospital, their placentas are burned after the birth. The placentas from women who birth at home are typically buried. Sometimes the placenta is buried in the corner of the yard, but really it does not matter where, according to one traditional healer. The placenta should be placed lightly in the hole, not tossed. It should be treated like a person. If it is thrown in the hole carelessly, the woman will have problems with her womb, he says. Others place the placenta in holes found in the ground in the forest. *Cohune palms* (*Attalea cohune* Mart.) are a ubiquitous part of the landscape in southern Belize, and unlike woody trees, when they die, these palms decay quickly, leaving holes around 1–2 ft in diameter, in the forest. On a plant collection trip, I asked my interpreter what was making all these large holes in the forest floor. She explained that those were left behind after cohune palms died and decayed, and that those holes were where her father would bury her mother's placentas after she gave birth to her and her siblings.

Hemorrhaging

Don Manuel uses *ichaj pim* (*hikalit kak*; *Psychotria acuminata* Benth.; JD 5) leaves to treat postpartum hemorrhaging (and heavy menstruation). A handful of leaves are boiled and the woman drinks just one cup if it works; she would drink more if she is still bleeding. This plant is considered a hot plant according to Don Manuel's system of classification. The plant is not common, but is found in both primary and secondary forests, he says.

For bleeding after childbirth, a handful of ***basil*** (*Ocimum campechianum* Mill.) leaves are boiled and given to the woman as a tea to drink twice, according to the male herbalist. To treat hemorrhage, three red roses (*Rosa* sp.) and a small amount of the leaves are boiled in a small amount of water and drank by the cup. Another treatment for hemorrhage calls for a handful of *ob'el* (cowfoot, *pata de vaca*: *Piper auritum* H.B.K.; JD 27) leaves to be boiled and six cups drank for the day. All the roots from a small *banana* (*Musa* x *paradisiaca* L.) plant, about 1 ft high, are boiled with 2–3 cups of water for 10 min to treat hemorrhage according to one herbalist. One tablespoon should be consumed every half an hour, but the woman should not have more than three doses, she says.

For one herbalist's remedy for postpartum hemorrhage (or heavy periods or miscarriage), "one dollar amounts" (approximately two tablespoons) of each of: *manzanilla* (thought to be *Matricaria recutita* L.) stems with leaves and flowers, *oregano* (thought to be *Lippia graveolens* H.B.K.) leaves, *chalche* (unidentified) leaves, along with *garlic* (*Allium sativum* L.) and three *cloves* (*Syzygium aromaticum* (L.) Merrill & Perry) are cut up. This is all boiled together in about a half-gallon of water until it turns dark brown, in about 10–15 min. This will make enough to drink three times for 1 day, and is consumed for 1 or 2 days. Another amount would be made for the second day if needed.

Roq wakax (unidentified plant, which has a cowfoot-shaped leaf, but is not the "cowfoot" *Piper auritum* also known as *ob'el*), can be used for hemorrhage, according to one midwife. The stem is cut into pieces and boiled in five cups of water. The woman would drink a cup three times a day. For hemorrhage, one healer would use *q'eq xeeb'* (unidentified) and *kaq uuqub'* (unidentified; JD 81) by mashing the leaves together, boiling and giving the tea to the patient to drink for 2 days, like water. This would be accompanied by massage, "knocking the feet," and banding the abdomen.

2.2.6 Breastfeeding and Weaning

Increasing Breast Milk

Many Q'eqchi' women drink warm *cacao* (*Theobroma cacao* L.) with *black pepper* (*Piper nigrum* L.) added to increase breast milk (Fig. 2.10). The stronger the *black pepper*, the more effective the drink is. A variation uses *pimienta gorda* (*Pimenta*

Fig. 2.10 Maya woman roasting cacao seeds to make cacao drink

dioica (L.) Merr.) seed ground and mixed with *cacao*. According to one midwife, *corn masa* (*Zea mays* L. subsp. *mays*) with warm water should also be consumed continuously to produce more breast milk. Another variation uses dried *corn* that is ground and roasted and made into a drink with *black pepper*, and consumed for 1 month. One traditional healer said breastfeeding women should have raw *pumpkin* (*Cucurbita* sp.) seeds mashed with water and mixed with *black pepper*. Another variation uses *sakil* (pumpkin: *Cucurbita* sp.) seeds toasted on the fire, ground and drank with water to increase breast milk. A topical treatment to "bring down breast milk," uses *san diego* (*yut u it*: *Piper peltatum* L.; JD 15) leaf to make a warm infusion to bathe the back and breasts with.

Don Francisco uses *rix saltul* (mamey: *Pouteria sapota* (Jacq.) H.E. Moore & Stearn) bark to increase women's breast milk production. He boils 4–5 in. of bark and has the women drink the liquid. He also uses *kuchim* (unidentified; JD 31) seed to make a remedy, has the woman eat the fruit, or he crushes the fruit in water and has the woman drink this. He prefers to use this over *cacao* (*Theobroma cacao* L.) because he says it works better. *B'ayl* (*Desmoncus orthacanthos* Mart.; JD 23) stems and *rix saltul* bark can be boiled for another recipe, or in conjunction with the *kuchim* treatment. Just *kuchim* and *b'ayl* could also be used together. He said to collect the "heart" of the young shoots of the *b'ayl*. The woman should drink the water after boiling this "heart" along with the *kuchim* seed. Another herbalist also suggested *b'ayl* to increase breast milk. She said the young soft leaves are boiled and eaten and the water is consumed like tea.

To increase breast milk, breastfeeding women should swallow seven *b'ulum* (small river fish approximately 1–2 in. long), according to Ms. Francisca. One herbalist says to collect nine little river fish and swallow them live and whole. Another said to "bring down breast milk," three small fish are swallowed alive. This practice was also reported among Garifuna healers interviewed in Southern Belize and Livingston, Guatemala. "The secret for more milk is for her to comb the sides of her waist with a hair comb," according to Don Manuel. Another herbalist agreed that combing the breasts in a downward motion with a hair comb helps to increase breast milk as well. *Coffee* (*Coffea arabica* L.) is thought to dry women's breast milk and often women who are trying to increase their milk supply will not drink coffee.

Weaning

One woman said she breastfed her children until she would get pregnant with another child and used *jackass bitters* (*Neurolaena lobata* (L.) R. Br. ex Cass.) rubbed on her breasts for weaning. The bitter leaves are crushed and rubbed on the breasts especially around the nipples, as a deterrent. Many Q'eqchi' women use this for weaning their children from breastfeeding. For weaning, Ms. Francisca recommends putting *hot pepper* (*Capsicum* sp.), *jackass bitters* (because of the hot or bitter tastes, respectively) or *achiote* (*Bixa orellana* L.) (because it colors the skin) on the breast to deter the child. One woman said she used *coffee* (*Coffea arabica* L.) topically on the breast for a deterring taste and *achiote* topically for a deterring look on the breasts.

2.2.7 Menopause

Causes of Menopausal Symptoms

Many Q'eqchi' feel that menopausal symptoms occur when women "do not take care of themselves," so healthy women should not have any bad symptoms. One traditional healer said that menopause is natural, so he does not know about treating this. Don Manuel explains that "menopausal hot flashes are caused sometimes because a woman is being forced to reach menopause because maybe during hospital deliveries she is given medications to stop getting pregnant so they need to go through a process with medicinal plants (to address this)." Contraceptives from doctors or "too many abortions" can cause difficult menopausal symptoms later, according to another traditional healer. He felt also that "women should use the rock during menstrual cycles" (presumably meaning rub a heated rock over their abdomen like after childbirth). It is believed that if a woman goes outside the first few days after giving birth the wind or cold could enter her, and could cause issues later during menopause.

Treatments for Menopausal Symptoms

Don Francisco knows several plants to treat menopausal symptoms and says they all have similar effectiveness. The accessibility of the plant determines which he decides to use. Two that he uses to treat hot flashes, sometimes in combination, are *b'aknel pim* (*Drymonia serrulata* (Jacq.) Mart.; JD 16) and *q'an pom* (*Piper jacquemontianum* Kunth; JD 3). Either can be used individually, or both together, since both are equally effective. Twenty leaves of *b'aknel pim* are crushed in cold water with 13 leaves of *q'an pom*. The woman drinks a small cup and bathes in the rest of the water. This would be done once a day for 3 days. These plants are considered cold remedies according to Don Francisco's classification system. He noted that *b'aknel pim* must be used fresh since it decays easily.

Don Francisco uses *q'an pom* (*Piper jacquemontianum* Kunth; JD 3) to treat menopausal hot flashes, to "cool them down." The plant's name means yellow incense because the plant has a scent that is similar to incense. *Q'an* means yellow, and *pom* is the word for incense. Twenty leaves are crushed into a small bucket of cold water. A cup of the liquid is given to the patient to drink, and the remaining liquid is used as a bath. This process should be done once per day every morning, but only for 4 days, when she is experiencing hot flashes. This remedy is only done with fresh leaves. This plant is considered a cold plant according to Don Francisco's system of classification. The plant is common and grows anywhere in primary or secondary forests he says.

Ik'b'o'lay pim (unidentified; JD 14) is another plant Don Francisco uses to treat menopausal hot flashes. *Ik'b'o'lay* is the name for the yellow jaw or fer de lance snake. This plant is named after a green snake called *rax ik'b'o'lay* that "puts itself on the ground very straight." *Rax* means green or blue. A handful or two of the leaves are crushed in a bucket of water. Then the woman would bathe with the water like normal bathing, only in the morning, every morning until she gets better. This plant can be stored for 2–3 months. It is considered a cold plant according to Don Francisco's classification system. He says the plant mostly flowers during the dry season, and gives a white flower. This is a common plant that grows in primary and secondary forests, according to him. Also, Don Francisco sometimes uses *xq'uq' mokooch* (unidentified) and *ik'b'o'lay pim* together for the woman to bathe in and wet her head with for menopausal hot flashes.

Don Manuel uses *pereen pim* (*Psychotria poeppigiana* Müll. Arg.; JD 24) to treat hot flashes, as the plant helps to "flow the problem out." He uses a tea of the boiled leaves for this treatment. Bitter plants like, *sorosi* (*serosi, yamor*: *Momordica charantia* L.; JD 69) are often used to treat menopausal symptoms. Also *g'an mank* (jackass bitters: *Neurolaena lobata* (L.) R. Br. ex Cass.), is used, but just 2–3 leaves because it is strong. For heat on the hands and feet, these two can be taken as a tea together until the woman is better, only twice a day in the morning and the evening. One herbalist says that the dried leaves of *dormilón* (sleeping prickle: *Mimosa pudica* L.; JD 50) can be smoked to relieve the symptoms of menopause. Another woman uses leaves of *chalche* (unidentified), oregano (*Lippia graveolens* H.B.K.; JD 67), *sorosi* (*serosi*: *Momordica charantia* L.; JD 69), and *epazote* (*Chenopodium ambrosioides* L.), boiled together and drank 2–3 times a day for difficulties in menopause. To treat menopausal hot flashes, one midwife uses massage and three plants, a handful of *may pim* (unidentified), *che'oreek* (oregano: *Lippia graveolens* H.B.K.; JD 19), and four leaves of *yut it puchuch* (unidentified) boiled in a pot of water. She should drink a warm cup of the tea three times a day for 3 days. (See *Lippia graveolens* in Fig. 2.11). One healer uses *kun che'* (unidentified) to treat menopausal hot flashes. The roots are boiled and given to the woman to drink hot for 3 days.

Fig. 2.11 *Che'oreek* (*Lippia graveolens* H.B.K.) leaves and flowers

2.2.8 Men's Reproductive Health

Male Fertility and Contraception

Ixq pim (*Tococa guianensis* Aubl.; JD 77), which means "female herb," is used as a male birth control, but this can cause permanent sterility which cannot be reversed, Don Francisco warns. He boils 13 leaves with the domatia (ant-housing structures, which are thought to resemble female human genitalia) in 1 L of water for 10–15 min. This makes three cups, but the man should only drink one. He says the 13 leaves correspond to 13 energy points in the body, 6 points on each side of the body (at the wrists, elbows, shoulders, hips, knees and ankles, and one on the center of the forehead). This is a hot remedy according to Don Francisco's system of classification. *Ixq pim* is effective as a male contraceptive, he says, until one drink of *teelom pim* (*Clidemia crenulata* Gleason; JD 94) tea is taken to reverse it.

According to Don Francisco, *teelom pim* (which means "male herb") can be used to treat male impotence. He would use 13 leaves, which also symbolize the 13 days it takes for the remedy to work, and some of the "balls" (domatia that resemble human testicles). One liter of water with 13 leaves with domatia boiled for 10–15 min makes three cups of tea. For males he would only give them one cup to drink. The plant can be stored dry for 1 or 2 months and still used. When asked why he thinks there is a clue on the plant (the "balls"), he said this is the way God made the plant, and he doesn't know why but his ancestors used these. He says these "balls" are the ants' home. This remedy will give the man strength so that he can be sexually active and have stamina. It is useful for treating men with "weak sperm." It is considered a hot remedy according to Don Francisco's classification system. The man cannot have anything cold during the treatment. Another remedy for impotent men is *cocolmeca* (*Dioscorea* sp.), which should be soaked for 10 days and one spoonful three times a day should be consumed, according to one herbalist.

There are plants to treat erectile dysfunction, and there are plants for when men's "sperm are not strong." When men's sperm are not strong, one healer uses the "limb" of *kun che'* (unidentified) mixed with the bark or skin of *xtz'ik laj ow* (unidentified, translates as "penis of the raccoon"), cut upwards. The two are mixed together, boiled, and consumed three times a day for 2 days. He says the cutting of *xtz'ik laj ow* must be done upwards not downwards because the man being treated "cannot pick up," (has erectile dysfunction). There are also plants in Q'eqchi' traditional medicine that are anaphrodisiacs, used to ease male sex drive, as well as ease female sex drive, according to the traditional healers.

Male-Specific Ailments

One culturally specific condition that is experienced primarily by males is *q'an yaj*, which is characterized by excess pain in the stomach around the navel. This occasionally occurs in women as well, though mostly afflicts adult men. *Q'an yaj* means yellow sickness according to one herbalist who says this mostly occurs in adults, rarely children and is felt as a pain around the navel that feels like a ball and if the person leans on a table against their stomach they can feel it jump. Some people die from this she said. She clarified that it is not a hernia. Another herbalist told me, "if the *obeah* man gives you, it could kill you." An *obeah* man refers to someone who practices malicious spells in this case. Don Francisco would treat *q'an yaj* with massage. He said this is caused by *solitario* (worms) or from missing meals. One healer uses *jalib'* (unidentified) roots, mashed and boiled until the water is yellow. The man would drink "one calabash full" of this (a calabash is the globe-shaped fruit of *Crescentia cujete* L. that is dried and hollowed out to make bowls and cups of varying sizes). One herbalist uses two barks, *k'a che'* (unidentified; thought to be *Acosmium panamense* (Benth.) Yakovlev) and *yaw* (unidentified; thought to be *Cedrela odorata* L.) drank as a tea. The man would drink this for 3 days, boiling the same pieces of bark in a half-gallon of water each day.

One healer uses a plant he calls *may q'ehen* (unidentified) to treat swollen testicles and a swollen lower body, which is caused by *obeah* (a malicious spell in this case) or if people accidentally pass in front of someone throwing something out that they used in *obeah* practice. The plant is cooked and drank to treat the swelling. He says the doctor cannot treat this. Men sometimes suffer from *k'uub'sa'* as well according to Don Francisco, which is considered something primarily women suffer from (as a disorder of the womb), but men just get diarrhea he says.

2.2.9 *Infant Health*

Birth and Postpartum Period

Anise (*Pimpinella anisum* L.) seeds are boiled tied in a cloth with *garlic* (*Allium sativum* L.), mashed, then strained and given to babies by some Q'eqchi' mothers during their first few days after they are born, in a small teaspoon amount, starting before their first stool to clean out their system. This treatment cleans the baby's body of anything, like fluids that the baby swallowed from the birth. To dry a baby's cord, it should be cleaned with alcohol, then have a warm cloth applied to help dry it and then have "sweet oil" applied. For the first few weeks, a cloth warmed on the *comal* is used to warm the baby's navel, belly, bottom and head, and their arms and legs should be tied down so they do not frighten themselves.

To protect newborn babies, a clove of *garlic* (*Allium sativum* L.) can be pinned in their clothes. They can be bathed with *marigold* (*Tagetes* sp.) every day for protection. A red bracelet, a red string tied around the wrist, red strings through babies' pierced ears, or anything red worn by the baby is used to protect them against spirits. All this is done to protect the baby from evil spirits and spiritual afflictions. Not only do people believe this will ward off evil, but some also believe that tying a red string around the baby's wrist will protect the baby from pregnant women, who are considered in a hot state during their pregnancy. If a child does not have this protection and sees a pregnant woman, this can cause inadvertent harm to the baby and cause them to have green stool and become ill.

Baby Care

The length of time of breastfeeding does not seem to have changed much in the last few generations. Women breastfeed for a length of time lasting in the range of 9 months to 4 years, with 2 years being average from my study. Some women said they stopped because the child stopped breastfeeding themselves, others say they stopped when the milk dried, while others say they stopped because they were pregnant with the next child, so that determined the length. The oldest woman interviewed, who was 75 said she breastfed each for a year at which point she would be pregnant with another. The woman who breastfed the longest, 3–4 years, is the wife of a traditional healer; at 47 she was still breastfeeding her 2-year–old child, who came and lifted her mother's shirt to nurse during our interview. Breastfeeding in public is very common, around town, on public buses, and everywhere. There are signs to encourage breastfeeding and breastfeeding alone for babies up to 6 months, in the clinics, both in Punta Gorda Town and in the villages of Toledo.

For pink eye (conjunctivitis), breast milk is put in the eye as many times as possible, and/or the early morning dew from the leaves of plants, like *hibiscus* (*Hibiscus rosa-sinensis* L. var. *rosa-sinensis*) or *coco* (*Colocasia esculenta* (L.) Schott). If babies have a sore in their mouth, honey is wiped in the inside of the mouth. *Epazote* (*Chenopodium ambrosioides* L.) is used to treat worms. One handful of the leaves plus some seeds and two cloves of *garlic* (*Allium sativum* L.) are ground with a mortar and pestle. Warm water is added to this and three teaspoons are given to the child to drink, and the belly is massaged downwards. This is done three times a day. The worms will come out in the stool. There is the belief that children get this when they "have too many sweets." Symptoms of worms in children include no appetite, no energy, stomach pains, wanting to sleep on the floor, spitting a lot, and picking their nose and eating it.

For children that cry too much because of too much heat in the body, one healer would use *saq q'ehen* (unidentified) by bathing the child in it every morning for 3 days. This is also good for children and adults who are restless and cannot sleep because of too much heat in the body. There is the belief that children should not go near newborn animals because the children will get very thin. "They shrivel up and

Fig. 2.12 The author with young Q'eqchi' mother and her first baby holding the comb that we were not to use on his uncombed hair

you can see it on the child's behind," the healer explained. If this happens *k'an te'* (unidentified) is soaked overnight outside and the water is used to wash the child in the morning. You can see heat or steam come off of the child when you do this and this is the disease leaving the body, according to the healer.

There is a belief that you should not comb a baby's hair until their baby teeth come in. If the hair is combed before then, the baby's teeth are thought to come in spaced apart, like the teeth of a comb. In the villages, you will sometimes notice several month old babies who have never had their hair combed. A friend and Peace Corps volunteer wanted to take a photo of me with a young Q'eqchi' woman and her baby. The mother said sure, but said she needed something … for the baby … My friend took one look at the baby's messy hair, and assumed she meant a comb and brought one out from his room. We tried to comb the baby's hair for the photo and were urgently stopped by the mother and told that we could not comb his hair until his teeth came in (Fig. 2.12). She explained that she had only wanted a ball or something for the baby to hold in the photo.

2.2.10 Culturally Specific Infant Conditions

Fallen Fontanel

Fallen fontanel is referred to as "mole drop" in Belizean Kriol. One woman I interviewed who treats fallen fontanel only, no other ailments, calls this illness, *kub'enaq xha'*. The fontanel itself is called the *xha'* and *kub'enaq* means sunken, she explained. This illness occurs in babies from newborn age to 3 years old. It is not common, and just depends if the child has a fall. She says the *xha'* falls when the child has a bad fall. The soft spot of the baby's head will sink, then they do not want to nurse, they become dehydrated, and then get diarrhea. She treats this by holding the baby on her lap and "knocking the baby's feet," then massaging the stomach and pushing up the palate inside the baby's mouth with her thumb. The children recover from this in 1 day with her treatment. She learned by treating her own children. She started to treat this nearly 20 years ago. Words spread after she treated one baby, and she wanted to stop doing the treatments, but everyone was telling her to continue. She treats approximately six or seven babies per month for mole drop. She usually does not receive payment for this, but accepts if someone wants to offer something.

Another pronunciation for mole drop was *xkub'e xha'* and is said to be caused not only from falling, but also caused by holding babies up and tossing them. According to one herbalist, mole drop is serious because the baby will stop breastfeeding and have diarrhea, and the family would need to go to a traditional healer to get this treated. Another midwife massages the baby's body, lifts up the mouth palate, blows on the fontanel (or possibly sucks—it was difficult to tell), *knocks* the feet three times, blows (or sucks) on the fontanel again, then picks the baby up by the feet with the baby on her lap and gently shakes the baby in a downward direction three times. This process is thought to bring the fontanel back up. This is typically done after bathing the baby or child to treat for fallen fontanel.

Susto

Babies can get frightened from falling, frightened by another person, or frighten themselves while playing, and suffer from a sickness known as *susto* (Spanish) or *kaanil* (Q'eqchi') which means "fright." It is worse when someone is frightened from behind. They start to suffer from diarrhea and vomiting, do not eat, only sleep, and jump and stir in their sleep. Adults can also suffer from this, though it mostly affects babies and children. The dip in the base of the front of the person's neck will be moving too fast as will the pulse, according to one traditional healer. He would treat with *teelom q'ehen* (unidentified) and *su sik* (unidentified "long-leaved plant"), boiled together and drank. If the person is treated with medicinal plants and it does not work, a piece of hair from each person in the house would be cut then burned under the baby along with *garlic* (*Allium sativum* L.) leaf and a piece of thatch from each corner of the house. According to one herbalist, people with *susto* get dehydrated and get diarrhea that is a different color and is light and watery. Sometimes you can treat with *garlic* she says. A clove would be smashed and hot water added to it and drank. This is done two or three times. One man says he would burn nine prickles from a plant called *gut agash* (unidentified) three or four times to treat *susto*.

To treat this, the hair or clothing of the frightener (animal or person) must be burned under the afflicted person, or the spit or sweat from the frightener is rubbed on the afflicted person's forehead and stomach and massaged downward. If someone already has worms inside and then gets *susto* the person could "go crazy and go all over," according to one herbalist. Another herbalist told me, "as a traditional Q'eqchi' woman I use all my plants I have around for treating sickness in the family. I do not adapt to other people's customs. I practice a lot of Q'eqchi' customs like cutting the hair when someone is sick. Cutting the hair of the person who caused the "fright" and the person who was frightened, and burning the hair under the person sick from "fright." Traditions like this are being lost the quickest because people think it is a superstition."

Mal de Ojo

In Q'eqchi', *mal de ojo* (evil eye) or the equivalent is referred to as *kaq u* or "red eye." This can cause unintentional harm in young babies especially by admiring someone's baby too much, especially if you do not hold or touch the baby. Some people are believed to cause this more easily, such as pregnant women. According to one herbalist, if a pregnant woman sees a newborn baby and does not hold the baby, the baby will stool green, and have heavy breathing, because the spirit of the baby in the pregnant woman's womb is "heavy." If this happens the pregnant woman will have to come hold the baby or make the sign of the cross on the baby's forehead with her saliva. Another herbalist told me, "it is not good to say the baby is *so* beautiful to someone" because that could cause the baby to become ill. It is all right to admire, however not in excess.

Conclusions

There is great depth to the traditional knowledge remaining in the Q'eqchi' communities of Belize, as evidenced by the detailed ethnomedical information presented in this chapter. Further studies could expand on ethnomedical practices, beliefs, and treatments related to men's reproductive health as well as infant and child health. Some of the medicinal plants mentioned in this study have not yet been collected and identified to species, and there are likely additional species that the Q'eqchi' use in reproductive ethnomedicine that were not mentioned during the research for this study.

Traditional healing and herbalism practices can be very particular to the individual practitioner, and people's knowledge is always being enhanced by experience and exposure to others' knowledge. Traditional Q'eqchi' medicine is dynamic and continues to advance on the individual level as remedies, preparations, and treatments are refined, or newly discovered. Maya traditional medicine at the medical system level is continuing to make advancements as well, and research efforts such as this study make a contribution in support of this continued advancement. Plant-based remedies and other traditional treatments are supplying primary healthcare for much of the Maya world, much as they always have. It is only that with each successive generation there are far fewer practitioners and keepers of this important knowledge.

Though in-depth knowledge of traditional plant-based treatments still remains in the Belizean Q'eqchi' community, today this knowledge is mostly held by Q'eqchi' elders, with very few young Maya apprenticing. The detailed documentation of such knowledge can be utilized in educational efforts to train Q'eqchi' youth in traditional Maya medicine. The results of this study function as a reference to those practicing traditional medicine, those who wish to learn traditional Q'eqchi' medicine, and biomedical practitioners who wish to take an integrative approach or provide culturally competent care to their indigenous patients. Efforts to integrate traditional indigenous medicine and biomedicine in Belize and regionally can benefit as well from thorough ethnomedical documentation and associated bioactivity verification.

References

Alexiades MN (1996) Collecting ethnobotanical data: an introduction to basic concepts and techniques. In: Alexiades MN (ed) Selected guidelines for ethnobotanical research: a field manual. The New York Botanical Garden, Bronx

Amiguet VT, Arnason JT, Maquin P, Cal V, Vindas PS, Poveda L (2005) A consensus ethnobotany of the Q'eqchi' Maya of southern Belize. Econ Bot 59(1):29–42

Amiguet VT, Arnason JT, Maquin P, Cal V, Sánchez-Vindas P, Poveda Alvarez L (2006) A regression analysis of Q'eqchi' Maya medicinal plants from southern Belize. Econ Bot 60(1):24–38

Balick MJ, Nee MH, Atha DE (2000) Checklist of the vascular plants of Belize. NYBG Press, New York

Bourbonnais-Spear N, Awad R, Maquin P, Cal V, Sanchez Vindas P, Poveda L, Arnason JT (2005) Plant use by the Q'eqchi' Maya of Belize in ethnopsychiatry and neurological pathology. Econ Bot 59(4):326–336

Bourbonnais-Spear N, Awad R, Merali Z, Maquin P, Cal V, Arnason JT (2007) Ethnopharmacological investigation of plants used to treat *susto*, a folk illness. J Ethnopharmacol 109:380–387

Browner CH (1985a) Plants used for reproductive health in Oaxaca, Mexico. Econ Bot 39(4):482–504

Browner CH (1985b) Traditional techniques for diagnosis, treatment, and control of pregnancy in Cali, Colombia. In: Newman LF (ed) Women's medicine: a cross-cultural study of indigenous fertility regulation. Rutgers University Press, New Brunswick

Bussmann R, Glenn A (2010) Medicinal plants used in northern Peru for reproductive problems and female health. J Ethnobiol Ethnomed 6:30

Cheng W-Y, Kuo YH, Huang C-J (2007) Isolation and identification of novel estrogenic compounds in Yam Tuber (*Dioscorea alata* Cv. Tainung No. 2). J Agric Food Chem 55:7350–7358

Comerford SC (1996) Medicinal plants of two Mayan healers from San Andrés, Petén, Guatemala. Econ Bot 50(3):327–336

Cosminsky S (1994) Childbirth and change: a Guatemalan study. In: Carol P. MacCormack Ethnography of fertility and birth, 2nd edn. Prospect Heights, Ill. Waveland Press

Cosminsky S (2001) Midwives and menstrual regulation: a Guatemalan case study. In: Van de Walle E, Renne EP (eds) Regulating menstruation: beliefs, practices, interpretations. University of Chicago Press, Chicago

Doyle BJ, Frasor J, Bellows LE, Locklear TD, Perez A, Gomez-Luarito J, Mahady GB (2009) Estrogenic effects of herbal medicines from Costa Rica used for the management of menopausal symptoms. Menopause 16(4):748–755

Joe MM, Jayachitra J, Vijayapriya M (2009) Antimicrobial activity of some common spices against certain human pathogens. J Med Plants Res 3(11):1134–1136

Kahn H (2006) Seeing and being seen: the Q'eqchi' Maya of Livingston, Guatemala and beyond. University of Texas Press, Austin

Kang SC, Lee CM, Choi H, Lee JH, Oh JS, Kwak JH, Zee OP (2006) Evaluation of oriental medicinal herbs for estrogenic and antiproliferative activities. Phytotherapy Res 20:1017–1019

Kelmanson JE, Jäger AK, van Staden J (2000) Zulu medicinal plants with antibacterial activity. J Ethnopharmacol 69:241–246

Maquin A, Pow A, Caal E, Sho A, Caal F, Choc L, Baki M, Choc M, Teul M Sr, Teul M Jr, Chub T (Kekchi Healers Association) Treyvaud-Amiguet V, Arnason JT, Maquin P, Cal V, Sanchez-

Vindas P, Poveda Alvarez L (2005) Handbook of Kekchi medicinal plants of Belize. Viking First Nation Press, Ottawa

Michel J (2006) Medical ethnobotany of the Q'eqchi Maya: perceptions and botanical treatments related to women's health. Dissertation, University of Illinois

Ososki AL (2004) Ethnobotany of rural and urban Dominican Republic: medicinal plants, women, and health. Dissertation, City University of New York

Pesek T, Cal M, Cal V, Fini N, Minty C, Dunham P, Sanchez P, Poveda L, Arnason J (2006) Rapid ethnobotanical survey of the Maya mountains range in southern Belize, Central America: a pilot study. Trees Life J 1:10

Rojas MO, Collins S, Cal V, Caal F, Knight K, Arnason J, Poveda L, Sanchez-Vindas P, Pesek T (2010) Sustaining rainforest plants, people and global health: a model for learning from traditions in holistic health promotion and community based conservation as implemented by Q'eqchi' Maya healers, Maya Mountains, Belize. Sustainability 2:3383–3398

Sarker PC (1981) Methods used in induced abortion in Bangladesh: an anthropological perspective. Soc Sci Med 15B:483–487

Chapter 3
Use of Phytoestrogens in Q'eqchi' Maya Ethnomedicine

3.1 Traditional Maya Medicine

The Maya have a rich system of traditional medicine with an extensive ethnopharmacopoeia of plant-based treatments. Remedies have been passed down orally, from healer to apprentice, likely since the beginning of Maya civilization around 2000 BCE. Traditional Maya healers serve as the primary health-care providers, for many Maya people, especially in remote areas of Central America. Traditional Maya medicine has not maintained the same level of formal institutional study and global prestige that some other traditional medical systems have, such as Traditional Chinese Medicine or Ayurveda. In partnering with scientists, biological assays can be used to further validate the medicinal qualities of plants utilized in traditional medical systems. Traditional Maya healers can further validate the efficacy and safety of the medicinal plant species that they have found to be effective medicines in their healing practices; and in doing such, aid in returning prestige to these healing practices that were once held in such high regard in Central America.

3.2 Ethnomedically Directed Bioactivity Investigation

In identifying medicinal plants with therapeutic value, clues to bioactivity can be found by studying nature itself, such as, observing the ecological interactions of plants with other organisms. For example, resistance to herbivory or fungal attack, or allelopathic interactions can be indicative of plants with bioactive constituents that may also have medicinal applications. Other indicators of bioactivity include uses of plants, by animals, such as primates, to self-treat illness (Rodriguez et al. 1985; Huffman 1997, 2003).

Human cultures, through centuries of development of rich traditional medical systems, have identified and employed plants with medicinal effects that likely contain bioactive components. Agreement on medicinally effective species for particular health ailments among distinct and spatially separated traditional medical systems can aid in identifying the most biologically active species. Research on traditional medicinal uses of plants has yielded such important therapeutics as

J. De Gezelle, *Q'eqchi' Maya Reproductive Ethnomedicine*,
SpringerBriefs in Plant Science, DOI 10.1007/978-3-319-10744-8_3

digitalis, aspirin, quinine, codeine, and the chemotherapeutic agents vincristine and vinblastine. Despite many notable discoveries, less than 1 % of the estimated 420,000 flowering plant species in the world have been exhaustively screened for their chemical composition and associated medicinal value (Cox and Balick 1994; Balick personal communication 2008). Research on the use of traditional plant-based remedies has considerable potential for the identification of bioactive compounds in nature, which have particular promise for human safety and efficacy.

3.3 Phytoestrogens and Their Therapeutic Importance

Phytoestrogens are biologically-active, plant-based compounds that resemble mammalian estrogens in either structure or function. They represent a variety of chemical classes, including lignans, isoflavones, coumestans, anthraquinones, and saponins (Ososki and Kennelly 2003); and often a single plant will contain more than one class of phytoestrogen (Usui 2006).

Endogenous human estrogens (17β-estradiol, estrone, and estriol) have diverse functions in human health, including roles in reproduction, male and female fertility, cardiovascular health, cognition, bone density and the development of various cancers, namely mammary, ovarian, prostate, colorectal, and endometrial (Deroo and Korach 2006; Usui 2006). There are two structurally similar isoforms of estrogen receptors in the human body, estrogen receptor α (ERα) and estrogen receptor β (ERβ), which differ in tissue distribution and ligand affinity, but do have some overlap.

Phytoestrogens can bind to either one or both human estrogen receptors, (ERα and ERβ), and may act either as agonists, antagonists, or in some cases both. They have diverse therapeutic potential, with either estrogenic or antiestrogenic effects on the body. These compounds have therapeutic potential to treat or prevent a range of ailments, including infertility, reproductive cancers, and neurodegenerative disorders both in women and men (Deroo and Korach 2006). Similar to endogenous human estrogens, phytoestrogens show a range of tissue and receptor specificity, providing evidence for promising discoveries of phytoestrogens that can address a range of distinct estrogen-related conditions. Of particular interest in recent years is their potential as a nonsteroidal treatment for menopausal symptoms, in light of the health risks of hormone therapy.

3.4 Hormone Replacement Therapy (HRT) and the Women's Health Initiative

Two large-scale HRT studies by the Women's Health Initiative (WHI) involving 25,000 postmenopausal women were abruptly halted when increases in serious health conditions were observed in the women taking HRT. The Estrogen-Plus-Progestin

Study was halted in 2002 after researchers found that health risks outweighed the benefits, and the Estrogen-Alone study was abandoned shortly after in 2004 due to safety concerns (Hulley and Grady 2004). Combined estrogen-plus-progestin therapies have been associated with greater incidence of invasive breast cancer, increased node-positive breast cancer (cancer cells found in the lymph nodes making it more likely to recur and spread), and more deaths attributed to breast cancer (Chlebowski et al. 2010). Combined estrogen-plus-progestin also increased the risk of ischemic stroke (Wassertheil-Smoller et al. 2003), was associated with doubling the risk of venous thrombosis (Cushman et al. 2004), may increase the risk of coronary heart disease (Manson et al. 2003), increases the risk for probable dementia (Shumaker et al. 2003), was associated with greater brain atrophy (Resnick et al. 2009), and may also increase the risk of ovarian cancer (Anderson 2003). Three years after the estrogen-plus-progestin treatment was terminated; women who had received this treatment retained a greater risk for fatal and nonfatal malignancies (Heiss et al. 2008).

This resulted in decreased popularity of HRT for menopausal health (Ettinger et al. 2003), leaving an unmet need for estrogenic treatments and an increasing interest in phytoestrogen based therapies as a safe alternative. Attention has been given to soy (*Glycine max* (L.) Merr.), red clover (*Trifolium pratense* L.), and hops (*Humulus lupulus* L.) among others as sources of phytoestrogens with particular therapeutic potential. The diversity of plant-based estrogenic compounds, their tissue and receptor specificity, and conflicting data on some known phytoestrogens substantiate the need for further research on new sources and their bioactive constituents. Ethnomedical research on the use of traditional plant-based remedies for reproductive health can contribute to identifying phytoestrogens, which have particular promise for human safety and efficacy, since such plants are already in use by humans, and in most cases have been for decades or centuries.

3.5 Q'eqchi' Maya Reproductive Ethnomedicine

The Q'eqchi' Maya utilize plant-based medicines traditionally for reproductive health and fertility regulation, as well as for nearly all other health ailments. These plants may be effective due to the presence of hormone-mimicking compounds, such as phytoestrogens. Traditionally, women's reproductive health was the realm of female Q'eqchi' healers and midwives, but due to the paucity of practicing female healers in Belize, male healers have taken on the responsibility of treating women's reproductive health, by necessity. However, this is not traditional and not considered to be culturally appropriate by either Q'eqchi' men or women. Documenting medicinal plants used in Q'eqchi' traditional medicine for reproductive health and analyzing the associated bioactivity is an important component in preserving this branch of Maya traditional knowledge.

Joanna Michel (2006) conducted her doctoral research on medicinal plants used for women's health by the Q'eqchi' Maya of Livingston, Izabal, Guatemala. Michel found that the Q'eqchi' were using 48 medicinal plant species to treat conditions related to pregnancy, childbirth, menstruation, and menopause. Twenty of those species were evaluated in bioassays relevant to women's health, 12 in estrogen receptor binding assays, and 11 in serotonin receptor binding assays. Five species bound to ERα, with *Smilax domingensis* Willd. displaying the most significant binding activity, and six species bound to ERβ, with *Piper hispidum* Sw. showing the most significant binding activity. Serotonin receptor binding activity was found in most species that were evaluated, including *Cissampelos tropaeolifolia* DC., *Piper auritum* Kunth, *P. hispidum* Sw., *Hibiscus rosa-sinensis* L., and *Hyptis verticillata* Jacq. (Michel et al. 2007).

3.6 Methodology

3.6.1 Belize Indigenous Training Institute

Field research was carried out in the Toledo District of Southern Belize, in collaboration with the Belize Indigenous Training Institute (BITI) and the associated Q'eqchi' Healers Association (QHA). The traditional healers and coordinator of these organizations share with the author the interest in documenting their medicinal plant knowledge and further validating the efficacy of plant-based medicines used in traditional Q'eqchi' medicine with relevant biological assays. A research agreement was signed between the researcher and the BITI.

3.6.2 Ethnobotanical Research

Research in Belize was carried out between 2007 and 2011. Semi-structured interviews, forest and home garden walking interviews, photo elicitation, plant collections, and observations of remedy preparations and treatments (Alexiades 1996) were used to collect data with 21 Q'eqchi' traditional healers, midwives, and herbalists (15 women and 6 men). Interviews were undertaken only after prior informed consent was obtained. The Institutional Review Board (IRB) of The Graduate Center of the City University of New York approved all methodology (Appendix I), and renewed the IRB approval annually (IRB Number: 07-03-1231). Research, plant collection, and export permits were obtained through the Belize Forest Department and renewed annually.

3.6.3 Species Selection and Plant Collection

Plant species were selected for evaluation in the bioassay analysis based on specific criteria: (1) They were used to treat estrogen-related ailments, especially if the species were used to treat multiple different estrogen-related ailments; (2) They were used to treat menopausal symptoms specifically; (3) If estrogenic or antiestrogenic bioactivity was already found in the same family, or preferably genus; (4) The number of healers and herbalists who used the species; (5) The knowledge level of individual(s) who used the species; and (6) The feasibility of bulk collection. Ten species were selected to be assayed for estrogenic activity, namely: *Clidemia crenulata* Gleason (JD 94), *Drymonia serrulata* (Jacq.) Mart. (JD 16), *Gouania lupuloides* (L.) Urb. (JD 4), *Miconia oinochrophylla* Donn. Sm. (JD 78), *Mimosa pudica* L. (JD 50), *Piper jacquemontianum* Kunth (JD 3), *Piper peltatum* L. (JD 15), *Psychotria acuminata* Benth. (JD 5), *Psychotria poeppigiana* Müll. Arg. (JD 24), and *Tococa guianensis* Aubl. (JD 77).

3.6.4 Memorandum of Agreement for Laboratory Research

Evaluation of Q'eqchi' medicinal plants for bioactivity was carried out after a Memorandum of Agreement (MOA) was signed by the Belize Indigenous Training Institute (BITI) and the University of Illinois at Chicago (UIC).

3.6.5 Medicinal Plant Extraction

Methanol extractions were made from leaves of all ten species, stems of *G. lupuloides*, and roots of *M. pudica* and *P. peltatum*. Dried tissue was ground using an electric grinding mill (Cuisinart Grind Central Coffee Grinder Model DCG-12BC), and 500 mg samples were extracted in 5 ml of 80 % methanol by sonication for 30 min. The resulting extract was centrifuged at 3500 rpm for 5 min and the supernatant was poured into pre-weighed vials. This method was repeated three times to ensure exhaustive extraction. Finally, extracts were dried under nitrogen gas, and lyophilized to remove all residual water.

3.6.6 UIC/NIH Center for Botanical Dietary Supplements Research

Estrogen and antiestrogen bioassays were carried out at the UIC/NIH Center for Botanical Dietary Supplements Research in the lab of Dr. Gail Mahady, of the Department of Pharmacy Practice at the University of Illinois at Chicago (UIC). Extracts were prepared and dried in New York in the lab of Dr. Edward Kennelly in the Department of Biological Sciences at Lehman College and shipped to UIC for evaluation.

3.6.7 Bioassay Materials

Water-soluble 17β-estradiol and the antiestrogen ICI-182,780 were obtained from Sigma Chemical Co (St. Louis, MO, USA). Fetal bovine serum (FBS), phenol red-free DMEM medium (with 1000 mg/l glucose and with L glutamine) and OPTI-MEM were purchased from Invitrogen (Carlsbad, CA, USA). Lipofectamine-2000 reagent was purchased from Invitrogen (Carlsbad, CA). The MTS Cell Titer Cell Proliferation Assay was purchased from Promega (Madison, WI, USA). The Mercury Pathway Profiling System 2 containing vectors pTA-SEAP2 control vector and pERE-TA-SEAP reporter vector and the Great EscAPeTM SEAP Chemiluminescence High-Throughput Detection Kit were purchased from Clontech (Palo Alto, CA, USA).

3.6.8 Cell Culture

MCF-7 (HTB-22) human breast cancer cells were obtained from American Type Culture Collection, Manassas, VA. The cells were maintained in Dulbecco's modified Eagle's medium containing 10% FBS and 100 U/ml penicillin and 100 μg/ml streptomycin in 75 cm flasks. Forty-eight hours prior to transfection, the cells were maintained in phenol red free DMEM medium supplemented with 0.5% FCS, in a humidified atmosphere with 5% CO_2 at 37 °C. For transfection, 0.5×10^5 cells were seeded per well of a 96-well plate in OPTI-MEM supplemented with dextran-charcoal stripped FBS. After 4 h, serum concentration was reduced to 1%. Prior to transfection, cells were washed with DMEM and incubated in phenol red-free OPTI-MEM without FBS as described by Nozawa et al. (2005).

3.6.9 Transient Transfections and Reporter Gene Assays

MCF-7 cells were transfected using 0.5 μl Lipofectamine 2000, and 0.1 μg plasmid DNA (SEAP vectors) per well of a 96-well plate. For determination of secreted alkaline phosphatase (SEAP) activity induced through an estrogen response element (ERE), the cells were transfected with the reporter vector pERE-TA-SEAP, coding for the SEAP gene (Cullen and Malim 1995; Treeck et al. 2008). As a negative control, reporter vector pTA-SEAP2 using Lipofectamine 2000 according to the manufacturer's instructions. After 12 h of transfection, the media was replaced with phenol-free Dulbecco's modified Eagle's medium containing 10% dextran-coated charcoal stripped FBS. The extracts were added to each well at 20 μg/ml, with or without Estradiol (E2). E2 was used as the positive control and Dimethyl sulfoxide (DMSO) as the negative control. Cells were treated for 12 h, and the supernatants were harvested, centrifuged and 25 μl from each well was placed into a new 96 well plate. The SEAP activity was measured using the Great EscAPeTM SEAP Chemiluminescence High-Throughput Detection Kit using a 96-well luminometer according to the manufacturer's instructions.

3.6.10 Cell Proliferation Assay

MCF-7 cells were plated at 10,000 cells/well in 96-well plates in phenol red-free medium containing 10 % DCC-FBS. After 24 h the cells were treated with 10 nM E2 or the medicinal plant extracts, or DMSO 0.1 % as a negative control. At 72 h after seeding, the cells were treated with trypsin 0.05 %/EDTA, and then the cell numbers were counted.

For the first SEAP activity reporter gene experiment, the extracts were assayed in triplicate and the last two experiments were repeated in duplicate for a total $N=7$. E2, an endogenous human estrogen, and ICI, an antiestrogen were used as positive controls and DMSO was used for a negative control. In the estrogen bioassay, treatments contained plant extracts alone. In the antiestrogen assay, each treatment contained a plant extract plus E2.

3.7 Results and Discussion

3.7.1 Cytotoxicity Results

The methanol extracts of leaves of *P. jacquemontianum*, and roots of *M. pudica* were both cytotoxic to the MCF-7 human breast cancer cells. We were therefore unable to assay them for estrogenic or antiestrogenic activity, although they remain of interest for their potential use in breast cancer therapy.

3.7.2 Estrogenic and Antiestrogenic Activities

Some estrogen-related health conditions can be related to estrogen deficiency (such as menopausal symptoms, amenorrhea, and infertility) and some can be related to estrogen excess or dominance (such as heavy and painful menstruation, some reproductive cancers, and infertility as well). Plants that stimulate the expression of the estrogen-responsive reporter gene in cell culture may provide useful treatments for estrogen-deficiency and the related conditions. Antiestrogenic plants that reduced estrogen responsive reporter gene expression in the presence of E2, may be of use in treatment of estrogen-excess or dominance and related conditions.

3.7.3 Fabaceae: Mimosoideae

Mimosa pudica L. is known as *wara k'ix* in Q'eqchi' Mayan or sleeping prickle in Belize Kriol (Fig. 3.1). The leaves and roots are used to treat menopausal symptoms and uterine fibroids in Q'eqchi' reproductive ethnomedicine. The dried leaves are smoked to relieve the symptoms of menopause. To treat "fibers in the womb"

Fig. 3.1 *Mimosa pudica* leaf and flower

(thought to be uterine fibroids) *M. pudica* is boiled with two other species and the tea given to the woman to drink daily. The extract of the leaves was both estrogenic and antiestrogenic, while the extract of the roots was cytotoxic to the MCF-7 breast cancer cells used in our assay.

The plant is native to Central and South American, but has been introduced in many other areas and grows as a weed now pantropically, and is considered invasive in some areas outside of its native range. The plant is used for childbirth, to shorten labor and remove the placenta in Trinidad and Tobago (Lans 2007), and traditionally used as birth control in parts of rural India (Ganguly et al. 2007). Methanol extract of the root was administered to mice for 21 consecutive days, which reduced their number of litters, and disturbed the secretion of gonadotropin and E2 and suppressed follicle-stimulating hormone (FSH). These prolonged the estrous cycle and disturbed ovulation, having an antifertility effect on the mice (Ganguly et al. 2007). *M. pudica* root powder was also administered to rats, which resulted in a significant reduction in the number of normal ova and an increase in the number of degenerated ova, compared to control rats (Valsala and Karpagaganapathy 2002).

3.7.4 Gesneriaceae

Drymonia serrulata Jacq. (Mart.) is known in Q'eqchi' as *b'aknel pim* and is used to treat menopausal hot flashes, menstrual pain, as well as pain and mild fever in pregnant women. The leaves are crushed in cold water, and then warm water is added

Fig. 3.2 *Tococa guianensis* flowers and leaves with domatia resembling female human genitalia

and given to women to drink and bathe with. The extract of the leaves was both estrogenic and antiestrogenic in our assays. The plant is also used in the traditional medicine of the Guaymi indigenous people and was evaluated for its analgesic and anti-inflammatory effects. An aqueous fraction of *D. serrulata* was shown to have peripheral analgesic activity as well as dose-dependent anti-inflammatory activity at 24 and 96 h (Espósito et al. 1993).

3.7.5 Melastomataceae

Clidemia crenulata Gleason is referred to as *teelom pim* (male herb) and is used in Q'eqchi' traditional medicine to treat male impotence and used as a female contraceptive. The plant is believed to have opposite effects on males and females who drink the tea. Couples will often use this plant in conjunction with *Tococa guianensis* Aubl., known as *ixq pim* (female herb) in Q'eqchi' Mayan, which is used conversely to treat female infertility and used as a male contraceptive. If the couple would like to conceive the male would drink the tea made from *C. crenulata*, and the woman would drink the tea made from *T. guianensis*. If the couple would like to decrease chances of conception, the couple would switch which plant species they consume as tea.

Both species are myrmecophites, which are plants that house ants in structures known as domatia. Myrmecophites are particularly diverse among the Melastomataceae family. Ants inhabit and nest in the domatia and may in turn protect the plants from potential herbivores (Michelangeli 2010). Leaves and domatia are used to prepare the teas used as contraception, infertility and impotence treatments.

In these two species their domatia resemble human genitalia, giving the plants their Q'eqchi' names meaning "female herb" (Fig. 3.2) and "male herb" (Fig. 3.3). The Doctrine of Signatures refers to the human use of medicinal plants based on some physical characteristic of the plant that resembles a human body part or human body characteristic, with the belief that these plant characteristics are suggestive of the plant's medicinal application. Scientists and skeptics often assume that plants used medicinally based on the Doctrine of Signatures must not have biologically

Fig. 3.3 *Clidemia crenulata* domatia resembling male human genitalia

active compounds and are only used because the resembling characteristic of the plant rather than actual observed efficacy. Contrary to these skeptical presumptions, the extract of the leaves of *C. crenulata* was estrogenic and the extract of the leaves of *T. guianensis* was both estrogenic and highly antiestrogenic in our assays.

A related species found in Guatemala, *Clidemia setosa*, bears setose domatia at the apices of its petioles that resemble those of *T. guianensis* in shape and placement, and is also used medicinally to treat sterility in women. It is known as *ixquiquen* in Verapaz, Guatemala. In *The Flora of Guatemala* (1963), *Clidemia setosa* is said to be "a fine example of the doctrine of signatures in practice of domestic medicine, since a decoction of the leaves, sometimes prepared almost ceremonially, is a favorite remedy for sterility in women. Dieseldorff remarks that its effect is doubtful, but in Guatemala perhaps as much faith is placed in it as in a pilgrimage to Esquipulas." In Oaxaca, Mexico *Clidemia setosa*, "herb of the female person" was reported to be "used to evoke the conception of a female child based on the Doctrine of Signatures," (Browner 1985) based on the belief that "plants that could produce and differentiate plant tissue into female and male forms could do the same for human genital tissue."

Miconia oinochrophylla Donn. Sm. is known as *xoy pim* in Q'eqchi' Mayan (Fig. 3.4). The leaves are used to treat heavy menstruation, menstruation that is too frequent, and to prevent miscarriages. The abaxial surface of the leaves are purple to fuchsia colored, and is likely another example of the Doctrine of Signatures. The leaves are crushed in cold or hot water and drank to treat heavy menstruation (Maquin et al. 2005). The leaves are crushed in cold water and given to a woman to drink for menstruation that is too frequent, or if she has had a history of miscarriages. The extract of the leaves of this species was the most estrogenic of the 13 extracts assayed in our study.

Fig. 3.4 *Miconia oinochrophylla* leaves showing fuchsia colored abaxial leaf surface

Fig. 3.5 *Piper jacquemontianum* leaves and inflorescences

3.7.6 Piperaceae

Piper jacquemontianum Kunth is known as *q'an pom* in Q'eqchi' Mayan (Fig. 3.5). The leaves are crushed in cold water which is then given to the woman to drink and bathe with to treat menopausal hot flashes, in Q'eqchi' reproductive ethnomedicine. The extract of the leaves did not display estrogenic activity, but was cytotoxic to the MCF-7 breast cancer cells used in the assay. An ethanol extract of this species also previously showed significant cytotoxicity against MCF-7 breast cancer, H-460 lung cancer, and SF-268 central nervous system cancer cell lines (Calderón et al. 2006).

Fig. 3.6 *Piper peltatum* leaves and inflorescences

Piper peltatum L. (syn: *Pothomorphe peltata*) is known as *yut it* or *yut u it* in Q'eqchi' Mayan (Fig. 3.6). The leaves are used as a galactagogue (to promote lactation), and to treat heavy menstruation, dysmenorrhea (painful menstruation), amenorrhea (lack of menstruation), as well as *k'uub'sa'* or "womb disorder," and "cold in the womb" which are culturally specific diseases in Q'eqchi' reproductive ethnomedicine. To promote lactation, a warm infusion is made from the leaves and used to bathe the back and breasts. To treat menstrual pain, the leaves are wrapped around a warmed rock and rubbed on the stomach. To treat heavy menstruation, the leaves are crushed in water and drank and used to wet the abdomen (Maquin et al. 2005). To treat amenorrhea the leaves and roots are boiled with other species and drank as a tea. The whole plant, including roots, stem, and leaves are boiled and drank as a tea to treat *k'uub'sa'*. This remedy is accompanied by traditional Q'eqchi' abdominal massage to return the womb to its proper position. The extract of the leaves was both an estrogen receptor agonist and antagonist, in our assays, and the extract of the roots was an estrogen receptor agonist.

Other *Piper* species have been shown to exhibit bioactivity or efficacy related to their uses as medicines for reproductive health. *P. hispidum*, used in Q'eqchi' ethnomedicine in Guatemala for dysmenorrhea and amenorrhea was shown to have a strong binding preference for ERβ, compared to other species that displayed similar binding affinity for both estrogen receptors, suggesting that the species may contain compounds that act as selective estrogen receptor modulators, also known as SERMS (Michel et al. 2007). *Piper guineense* is used in Nigeria to promote conception, as well as to stimulate uterine contractions after childbirth to expel the placenta. This species was shown to enhance rat uterine contractions and increase uterine weight similar to E2, and therefore concluded to have both estrogenic and oxytocic properties (Udoh 1999).

Fig. 3.7 *Gouania lupuloides* stems and leaves

3.7.7 *Rhamnaceae*

Gouania lupuloides (L.) Urb. is known as *ch'ajom k'aham* in Q'eqchi' Mayan, which means "young man vine" (Fig. 3.7). The plant is used to treat heavy menstruation in Q'eqchi' reproductive ethnomedicine. The leaves or stems are boiled and drank as a tea. Both the extract of the leaves and the extract of the stems were estrogenic in our assay, though the stems were slightly less estrogenic than the leaves. *G. lupuloides* was also reported to be used by Latino healers in New York City for the treatment of uterine fibroids (Balick et al. 2000). This species is known as *chewstick* in Jamaica and other parts of the West Indies, where it is used for oral hygiene. The stick is bitter tasting and becomes frothy when the end of the stick is chewed. The plant is known to contain triterpenoid saponins (Kennelly et al. 1993).

3.7.8 *Rubiaceae*

Psychotria acuminata Benth. is known as *ichaj pim* in Q'eqchi' Mayan and is used to treat heavy menstruation and postpartum hemorrhage in Q'eqchi' reproductive ethnomedicine (Fig. 3.8). The abaxial leaf surfaces of this species are also fuchsia to purple colored, and is likely another example of the Doctrine of Signatures, considering the uses. The leaves are crushed in cold water or boiled, and given to the woman to drink. The extract of the leaves was an estrogen receptor agonist in our assay.

Psychotria poeppigiana Müll. Arg. is known as *pereen pim* in Q'eqchi' Mayan and hot lips in Kriol because of the characteristic bright-red bracts (Fig. 3.9). The leaves are used to treat menopausal hot flashes and amenorrhea (lack of menstruation) in Q'eqchi' reproductive ethnomedicine. The leaves are boiled and drank as a tea. The extract of the leaves was an estrogen receptor agonist in our assay.

Fig. 3.8 *Psychotria acuminata* flowers and leaves showing color on abaxial leaf surface

Fig. 3.9 *Psychotria poeppigiana* leaves and bracts

Conclusion

Findings from this research contribute to the literature on Q'eqchi' Maya ethnomedicine by associating the biological activity with the traditional remedies, subsequently further validating traditional Maya medicine and broadening the understanding of the efficacy of these plants. Documentation of traditional medical treatments aids in preserving the knowledge of these traditions and creates documents that can be used in the training of new apprentices in traditional Maya medicine, in addition to the long tradition of studying one on one with an elder traditional healer. We hope that in demonstrating the associated bioactivity in the medicines that are being used in this traditional medical system, these results will further validate and return prestige to traditional Maya medicine.

Q'eqchi' Maya reproductive ethnomedicine is using phytoestrogens to treat many reproductive health ailments including menopause, menstrual difficulties, and infertility. Biomedical systems have much to learn from the ancient and contemporary wisdom of traditional medical systems. A meta-analysis of 174 studies using randomized controlled trials, found that phytoestrogen supplements have a safe side-effect profile with only moderately elevated rates of gastrointestinal side effects (Tempfer et al. 2009). In addition to contributing to the validation of traditional Maya medicine, this work also contributes to the data regarding plant sources for hormone-mimicking medicines, which may be safe and efficacious for humans.

References

Alexiades MN (1996) Collecting ethnobotanical data: an introduction to basic concepts and techniques. In: Alexiades MN (ed) Selected guidelines for ethnobotanical research: a field manual. The New York Botanical Garden, Bronx, pp 53–94

Anderson EN (2003) Those who bring the flowers: Maya ethnobotany in Quintana Roo, Mexico. ECOSUR: Chetumal, Quintana Roo

Balick MJ, Kronenberg F, Ososki AL, Reiff M, Fugh-Berman A, O'Connor B, Roble M, Lohr P, Atha D (2000) Medicinal plants used by Latino Healers for women's health conditions in New York City. Econ Bot 54(3):344–357

Browner CH (1985) Plants used for reproductive health in Oaxaca, Mexico. Econ Bot 39(4):482–504

Calderón AI, Vásquez Y, Solís PN, Caballero-George C, Zacchino S, Gimenez A, Pinzón R, Cáceres A, Tamayo G, Correa M, Gupta MP (2006) Screening of Latin American plants for cytotoxic activity. Pharmaceut Biol 44(2):130–140

Chlebowski RT, Anderson GL, Gass M, Lane DS, Aragaki AK, Kuller LH, Manson JE, Stefanick ML, Ockene J, Sarto GE, Johnson KC, Wactawski-Wende J, Ravdin PM, Schenken R, Hendrix SL, Rajkovic A, Rohan TE, Yasmeen S, Prentice RL, WHI Investigators (2010) Estrogen plus progestin and breast cancer incidence and mortality in postmenopausal women JAMA 304(15):1684–1692

Cox PA, Balick MJ (1994) The ethnobotanical approach to drug discovery. Sci Am 270(6):82–87

Cullen BR, Malim MH (1995) Secreted placental alkaline phosphatase as a eukaryotic reporter gene. Meth Enzymol 216:362–368

Cushman M, Kuller LH, Prentice R, Rodabough RJ, Psaty BM, Stafford RS, Sidney S, Rosendaal FR, Women's Health Initiative Investigators (2004) Estrogen plus progestin and risk of venous thrombosis. JAMA 292(13):1573–1580

Deroo BJ, Korach KS (2006) Estrogen receptors and human disease. J Clin Invest 116(3):561–570

Espósito AM, Gupta MP, Calderón A, Zamora VO, Buitrago de Tello R (1993) (Spanish) The analgesic and anti-inflammatory effects of *Drymonia serrulata* (Jacq.) Mart. Rev Med Panama 18(3):211–216

Ettinger B, Grady D, Tosteson ANA, Pressman A, Macer JL (2003) Effect of the women's health initiative on women's decisions to discontinue postmenopausal hormone therapy. Obstet Gynecol 102(6):1226–1232

Ganguly M, Devi N, Mahanta R, Borthakur MK (2007) Effect of *Mimosa pudica* root extract on vaginal estrous and serum hormones for screening of antifertility activity in albino mice. Contraception 76(6):482–485

Heiss G, Wallace R, Anderson GL, Aragaki A, Beresford SAA, Brzyski R, Chlebowski RT, Gass M, LaCroix A, Manson JE, Prentice RL, Rossouw J, Stefanick ML, WHI Investigators (2008) Health risks and benefits 3 years after stopping randomized treatment with estrogen and progestin. JAMA 299(9):1036–1045

Huffman MA (1997) Current evidence for self-medication in primates: a multidisciplinary perspective. Yearbook Phys Anthropol 40:171–200

Huffman MA (2003) Animal self-medication and ethno-medicine: exploration and exploitation of the medicinal properties of plants. Proc Nutr Soc 62:371–381

Hulley SB, Grady D (2004) The WHI estrogen-alone trial—do things look any better? JAMA 291(14):1769–1771

Kennelly EJ, Lewis WH, Winter REK, Johnson S, Elvin-Lewis M, Gossling J (1993) Triterpenoid saponins from *Gouania lupuloides*. J Nat Prod 56(3):402–410

Lans C (2007) Ethnomedicines used in Trinidad and Tobago for reproductive problems. J Ethnobiol Ethnomed 3:13

Manson JE, Hsia J, Johnson KC, Rossouw JE, Assaf AR, Lasser NL, Tervisan M, Black HR, Heckbert SR, Detrano R, Strickland OL, Wong ND, Crouse JR, Stein E, Cushman M, Women's

Health Initiative Investigators (2003) Estrogen plus progestin and the risk of coronary heart disease. N Engl J Med 349(6):523–534

Maquin A, Pow A, Caal E, Sho A, Caal F, Choc L, Baki M, Choc M, Teul Sr M, Teul Jr M, Chub T (Kekchi Healers Association), Treyvaud-Amiguet V, Arnason JT, Maquin P, Cal V, Sanchez-Vindas P, Poveda Alvarez L (2005) Handbook of kekchi medicinal plants of Belize. Viking First Nation, Ottawa

Michel J (2006) Medical ethnobotany of the Q'eqchi Maya: perceptions and botanical treatments related to women's health. University of Illinois at Chicago Doctoral Dissertation

Michel J, Duarte RE, Bolton JL, Huang Y, Caceres A, Veliz M, Soejarto DD, Mahady GB (2007) Medical potential of plants used by the Q'eqchi' Maya of Livingston, Guatemala for the treatment of women's health complaints. J Ethnopharmacol 114:92–101

Michelangeli FA (2010) Neotropical myrmecophilous melastomataceae: an annotated list and key. Proc Calif Acad Sci 61(9):409–449

Nozawa T, Suzuki M, Yabuuchi H, Irokawa M, Tsuji A, Tamai I (2005) Suppression of cell proliferation by inhibition of estrone-3-sulfate transporter in estrogen-dependent breast cancer cells. Pharmaceut Res 22:1634–1641

Ososki AL, Kennelly EJ (2003) Recent trends in phytoestrogen research. Phytotherapy Res 17:845–869

Resnick SM, Espeland MA, Jaramillo SA, Hirsch C, Stefanick MI, Murray MA, Ockene J, Davatzikos C, Women's Health Initiative Memory Study (2009) Postmenopausal hormone therapy and regional brain volumes: the WHIMS-MRI study. Neurology 72:135–142

Rodriguez E, Aregullin M, Nishida T, Uehara S, Wrangham R, Abramowski Z, Finlayson A, Towers GHN (1985) Thiarubrine A, a bioactive constituent of *Aspilia* (Asteraceae) consumed by wild chimpanzees. Experientia 41:419–420

Shumaker SA, Legault C, Rapp SR, Thal L, Wallace RB, Ockene JK, Hendrix SL, Jones BN, Assaf AR, Jackson RD, Kotchen JM, Wassertheil-Smoller S, Wactawski-Wendi J, WHIMS Investigators (2003) Estrogen plus progestin and the incidence of dementia and mild cognitive impairment in postmenopausal women. JAMA 289(20):2651–2662

Tempfer CB, Froese G, Heinze G, Bentz EK, Hefler LA, Huber JC (2009) Side effects of phytoestrogens: a meta-analysis of Randomized Trials. Am J Med 122:939–946

Treeck O, Weber A, Boester M, Porz S, Frey N, Klaus D, Ortmann O (2008) H-ras dependent estrogenic effects of epidermal growth factor in the estrogen-independent breast cancer cell line MDA-MB-231. Breast Cancer Res Treat 80:155–162

Udoh FV (1999) Uterine muscle reactivity to repeated administration and phytochemistry of the leaf and seed extracts of *Piper guineense*. Phytother Res 13:55–58

Usui T (2006) Pharmaceutical prospects of phytoestrogens. Endocrine J 53(1):7–20

Valsala S, Karpagaganapathy PR (2002) Effect of *Mimosa pudica* root powder on oestrous cycle and ovulation in cycling female albino rat, *Rattus norvegicus*. Phytother Res 16(2):190–192

Wassertheil-Smoller S, Hendrix SL, Limacher M, Heiss G, Kooperberg C, Baird A, Kotchen T, Curb JD, Black H, Rossouw JE, Aragaki A, Safford M, Stein E, Laowattana S, Mysiw WJ, WHI Investigators (2003) Effect of estrogen plus progestin on stroke in postmenopausal women. JAMA 289(20):2673–2684

Chapter 4
Q'eqchi' Women's Lives, Healthcare and Cultural Loss

4.1 Q'eqchi' Maya Women's Lives in Belize

Out of Belize's six districts, Toledo is the southernmost and is often referred to as the *forgotten district* because of its geographical distance from the capital (Crooks 1997), and perception that resources allocated to the district are limited, relative to other districts in Belize. Toledo has the largest indigenous population and the lowest population density, the largest average household size and the least access to electricity and public water in the country, as well as many other indicators of disadvantage among the Belizean population. Furthermore, Toledo has the greatest disparity between males and females with regard to education levels and employment rates (Statistical Institute of Belize 2007). The lives of Q'eqchi' Maya women in Toledo differ considerably based on factors including their age, education level, whether they live in one of the remote villages of the district or in the district town of Punta Gorda, and how traditional their family life was, among other variables. Q'eqchi' women often say that their greatest struggle is finding employment with limited English and employable skills, and their greatest hope for their daughters is that they will be able to obtain better education and have more career opportunities than they did themselves. However, likely by necessity, many Maya daughters spend considerable time helping their mothers with housework and caring for their younger siblings (Fig. 4.1).

In Q'eqchi' villages, there are gender-based divisions of labor, with women's work being mostly in the domestic realm. Women take care of cooking, cleaning, childcare, food processing, drying grain, tending to animals, and growing small crops around the house. Men maintain the *milpa* and other agricultural tasks, hunting, and forest-based activities far from the home. There are a number of other family tasks for which there are gender preferences, but there is some room for negotiation, choice, or sharing of responsibilities (Wilk 1997). Gender roles and gendered knowledge are taught to young Q'eqchi' boys and girls by their parents in part through the chores that young people carry out. According to Zarger (2002), "Girls learn to bake tortillas (*xorok*), wash clothes (*puchuk*) and dishes, sweep (*mesubk*), and prepare food, while boys learn to "chop" bush in preparation for *milpa* (*k'alek*), cut firewood (*tsibk*), plant (*awk*), and fish (*karabk*)" (Fig. 4.2). Both girls

J. De Gezelle, *Q'eqchi' Maya Reproductive Ethnomedicine*,
SpringerBriefs in Plant Science, DOI 10.1007/978-3-319-10744-8_4

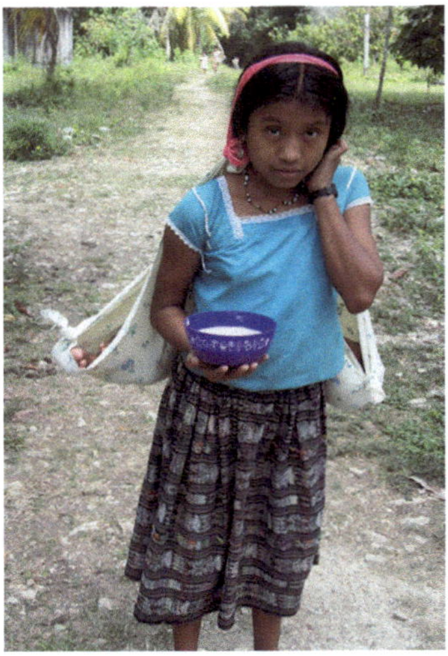

Fig. 4.1 Q'eqchi' girl carrying her baby brother in a *lepob'*

Fig. 4.2 Q'eqchi' boys returning to the village after chopping firewood

and boys are expected to find, identify, and harvest cultivated and wild food plants. Q'eqchi' youth typically learn these skills from family members of the same gender (Zarger 2002).

Early marriage, when girls are in their teenage years, is still fairly common in the Q'eqchi' communities, especially in the most remote villages. This means women have long reproductive lives, in some cases having 10–15 children. It is not uncommon for 20 years or more to separate the ages of a Q'eqchi' woman's youngest child and her oldest (Wilk 1997). According to Wilk (1997), living in multiple-family households can ease the burden of caring for so many children by spreading the workload out across the household. He explains that early marriage and high birth rates used to be balanced by an unfortunate high infant mortality rate, and that in the past, older children helped even more with the childcare. While infant mortality remains high it has decreased in recent decades. Furthermore, primary schooling has become mandatory in the villages of Belize. Wilk explains that the changes in these two factors—decrease in infant mortality and school becoming mandatory—has increased the burden on Q'eqchi' Maya mothers. The only way

Fig. 4.3 My Research Assistant Maria with her grandmother, mother, and daughter

to reduce this domestic workload is to share with other women, usually in familial groups or households.

An elder female herbalist who participated in this research was born to a mother from Belize, from the forested area outside one of the villages in Toledo, and a father from Guatemala. They were both Q'eqchi'. He was a soldier and ran from the service and came to Belize where he met her mother. She was their only child that survived past 7 months. Her other siblings died before that age. She lived in the forest there until she was 12 years old. At the age of 8 her mother died, and when she was 12 her father died as well. She had no schooling. Orphaned and alone, she moved out of the forest at the age of 12 to live with her uncle. Her uncle made her move in with a widower when she was 12 years old, so that his family would not have to take care of her. She is now 75 years old and has been widowed herself for the last 50 years, so has lived alone for a long time. She gave birth to 14 children, 6 are alive today. She has many grandchildren and great grandchildren. Her father and husband were both traditional healers, but both died while she was young. She herself had raised livestock, grown food to sell, and helped other people cook and wash to make money to support herself and her children.

Circumstances for many women of her generation may have been particularly difficult. Today life is still hard for many, while others find opportunity through education. Q'eqchi' women who live in Punta Gorda Town, or areas close by, are more likely to attend and complete high school, to speak fluent English comfortably, pursue an Associate's degree from the local campus of the University of Belize, and find employment, such as office work, teaching, and tourism, like my Research Assistant Maria (Fig. 4.3). In Belize, high school education requires that parents pay tuition and registration fees, and buy uniforms and books. In Toledo's remote areas, where many parents may live off of subsistence farming and have large families, parents may have to decide which children will be able to attend high school, and which will not, based on their limited budget. This decision often does not favor the female children, though, of course, every family is different.

4.2 Access and Choices in Reproductive Healthcare

Reproductive healthcare is limited in the Toledo District relative to the other districts, just as many other services are. Aminata Maraesa's doctoral study (2009) *I No "Fraid for That": Pregnancy, Risk, and Development in Southern Belize*, examined development initiatives related to reproductive health, the range of local maternity services and service providers from Belize's Ministry of Health (MOH) to village-based traditional healers and midwives, the reproductive realities of pregnant women in the Toledo District, and the interactions among all of these. In Toledo, she explains, "competing assessments of maternal risk, local interpretations of global health paradigms, and the cultural attitudes and beliefs surrounding women's bodies form the basis for reproductive decisions." Maraesa's work discusses the realities that women live during their reproductive years, which for development purposes are considered only as anonymous categories of reproductive health statistics, which international development programs are based upon.

Regarding Toledo, Maraesa (2009) explains that the Punta Gorda hospital provides basic midwifery services and medication to augment labor; however, the care is limited by the absence of an operating room and an on-site gynecologist. Therefore, all high-risk pregnancies and complicated labors must be sent to other hospitals out of the district for childbirth, although Punta Gorda hospital personnel indicated to Maraesa that emergency transfers were rare. Most women living in or close to Punta Gorda Town give birth successfully at the hospital in town. For those living in more remote areas of Toledo, Maraesa explains how:

> For those living in more remote areas of Toledo, Maraesa explains that women who live far from Punta Gorda give birth at home, mostly without any trained medical assistance, and mostly without complication.

Traditional birth attendants (T.B.A.s) have been trained in Belize by the MOH since the late 1950s, known locally as *nannies*. The Belize MOH offered 6 months of midwifery instruction to women who were known in their communities to already be knowledgeable in midwifery or providing related services. After these trainings, the MOH would recognize women as certified birth attendants and encourage these women to maintain a relationship with the MOH. The World Health Organization asserted that the training of T.B.A.s was a means to reduce maternal mortality rates in poor, developing countries, where childbirth outside of medical facilities is common. The T.B.A.s were to be trained under a standardized code that includes recognition of biomedically defined risk factors and referrals for high-risk pregnancies (Maraesa 2009).

Upon beginning my fieldwork, it became apparent to me that traditional indigenous midwives were no longer practicing in Belize. Maraesa (2009) also reported that she had not seen evidence of indigenous midwives or even Belizean *nannies*, who had received training from the MOH, currently practicing among the Mopan and Q'eqchi' speaking Maya in the Toledo District. Outside of the MOH hospitals and clinics, the indigenous healthcare providers attending to pregnant women were the male traditional healers, also known as "bush doctors" or *ilonels* in Q'eqchi' Mayan, but their services were usually limited to prenatal and postpartum treatments, and assistance in childbirth only during emergencies.

4.3 Women's Healing Traditions

My observations in Belize compared to the literature on traditional midwifery practices in Latin America led me to wonder, what had happened to traditional midwifery in Belize and more broadly to women's healing traditions, in the indigenous communities of Belize? And furthermore, what would a thriving contemporary practice of women's traditional healing be like in Belize? A look at indigenous women's healing traditions in the region may shed some light on this. Cosminsky (1994) explained that:

> In many societies, a specialist or midwife is a pivotal figure in the birth system and serves as both a medical and a ritual specialist. The role of the midwife and many traditional birth-related practices are being influenced and changed in a variety of ways, especially through the spread of Western medicine, both through directed changes, such as midwifery training programmes, and undirected changes, as through radio advertisements.

In Guatemala, Michel (2006) writes that "most Q'eqchi villages have at least one midwife who specializes mostly in pregnancy and childbirth, but also treats other gynecological conditions including menstruation, infections, and menopausal symptoms." She interviewed four Q'eqchi' midwives and one Garifuna midwife for her research, in Livingston, Izabal, Guatemala. The midwives utilize abdominal massages, administer hot herbal teas and poultices during and after birth, and treat menstrual and menopausal difficulties with plants that are cultivated or purchased at the market.

According to Cosminsky (1994), the indigenous Quiché Maya midwife in Guatemala, called an *iyom* or *ilonel* in Quiché and *comodrona* in Spanish, is recruited for this position traditionally through some sort of supernatural calling. She explains that:

> The destiny or *mandado* is revealed in various ways, such as through birth signs or omens, dreams, illness and finding strange objects (shells, scissors, special shaped stones and mirrors). The dreams and signs are interpreted by a diviner or shaman as signifying her destiny as a midwife, and the objects are considered messages sent by the supernatural (God or the spirits)…If a person does not follow one's calling, she shall suffer supernatural sanctions in the form of illness or death either to herself or members of her family.

Cosminsky (1994) explains, however, that in the Guatemalan Quiché Maya community, several of the midwives have received training by the Public Health Ministry. Between 1952 and 1978, the courses have consistently stressed the importance of hand washing and other infection control measures, horizontal delivery, and the recognition of complications that require referral to local doctors or hospitals.

Herbal medicine, use of the sweat bath, birthing in the squatting position, and other traditional practices were strongly discouraged. She goes on to add that "The official license and training is offered as an alternative to the divine mandate, opening up the role to others who want to practice." Ch'orti' Maya midwives in eastern Guatemala are also obliged to participate in training programs, and have been discouraged from using traditional practices including medicinal plants and external cephalic version (Kufer 2005).

Hinojosa (2004) writes of the highland Maya midwives who were retrained by Guatemalan health authorities. She contends that:

> Health officials are interested in midwifery only inasmuch as it can further their own objectives. To do this, health officials have enacted a system designed to cast in doubt the midwives' knowledge and gradually supplant it with biomedically sanctioned knowledge, knowledge with an authoritative basis that precludes other ways of knowing.

Of Costa Rica, Jenkins (2003) describes how this Central American country, in particular, is approaching the complete disappearance of traditional midwifery. This push for the total biomedicalization of birth began in the 1950s and 1960s. Jenkins explains that, "ultimately, the erosion of midwifery was initiated by top-down forces that first converted midwives into bridges to biomedicalization, then rendered their practices illegal because they were considered irrelevant to birth when viewed 'from above'."

In Belize, the MOH has been actively training T.B.A.s since 1957 in attempts to improve childbirth outcomes and reduce maternal fatalities throughout the country (Maraesa 2009). T.B.A.s are women or men in Belize, and in some cases, are also their village's Community Nurse's Aide (C.N.A.). A C.N.A. is chosen by the members of their village at a community meeting called by the *alcalde*, or village leader, and is then to be approved by the MOH. The C.N.A.s are trained in basic first aid, distributing over-the-counter medications, and measuring blood pressure, and meant to act as a liaison between the village and hospital (Maraesa 2009). Other women trained to be T.B.A.s had already been practicing as lay midwives or had some relevant experience or interest. It is expected that women in Punta Gorda Town will give birth at the hospital in town, and they do not have much of an alternative since the last homebirth midwife who lived in town passed away a decade ago. The MOH, however, acknowledges that this is difficult for women living in some of the more remote villages and they are urged to seek the assistance of a T.B.A., if possible (Maraesa 2009).

4.4 Methodology

4.4.1 Research Objectives

The objectives of the ethnographic portion of this research were the following: (1) To identify some of the factors affecting the lives and livelihoods of Q'eqchi' Maya women in the Toledo District of southern Belize; (2) To investigate the range of Q'eqchi' women's perspectives on reproductive health and healthcare; and (3) To understand why there has been a disproportionate decline in women's healing traditions in the Q'eqchi' Maya communities of Belize.

4.4.2 Interviews with Q'eqchi' Women

I interviewed ten Q'eqchi' Maya women in one of the villages along the Southern Highway in the Toledo District of southern Belize, during March 2008. All interviews were undertaken only after obtaining prior written informed consent provided in the participant's native language. Consent forms were reviewed and approved by the City University of New York's Institutional Review Board (IRB Number: 07-03-1231). These women were mostly from different families, known to my interpreter, throughout the village. They were not selected at random and the sample size was relatively small, but the aim was to get a snapshot of some of the range of experiences, practices, and perceptions of Q'eqchi' women living the village life, though along the highway with easy access to the district town of Punta Gorda, less than an hour away by bus that passes several times a day each way.

4.4.3 Interviews with Midwives, Herbalists and Healers

I interviewed 3 Q'eqchi' Maya midwives, 12 female herbalists, and 6 male healers in Toledo between the years 2007 and 2011, whose knowledge has contributed to this portion of the research as well. Occasionally, my Q'eqchi' research assistants also added the information, one female and one male.

4.5 Results and Discussion

4.5.1 Q'eqchi' Women of a Toledo Village

I interviewed women living in a village in Toledo, along the Southern Highway, with bus service several times daily to and from Punta Gorda Town. The village is typically transitional, with most multigenerational Maya households spanning a spectrum of traditional Maya to Maya living like your typical Belizean. Of the women interviewed, two were born in Guatemala, and eight were born in southern Belize—four were born there in the same village, and four from three other villages in Toledo. The ten women's age ranged from 23 to 56, with a mean age of 39 years. Seven of the ten women were married, one widowed, and two unmarried. All ten women had children, ranging from 1 to 10 children each, with an average of five children. The three oldest women each had the most children, both women in their

50s had ten children, and one woman in her late 40s had eight children. Those three women were each married between the ages of 14 and 16. Many of the women also grew up in families in which they were one of the 8–10 siblings.

Schooling ranged from no schooling (two women), to a few months in Guatemala (one woman), to some primary school (two women), to completed primary school (four women), to completed secondary school (one woman). All ten women could speak Q'eqchi' Mayan, four could also speak Belizean Kriol/English, three could speak or understand a little Belizean Kriol/English, one could speak a little Mopan Mayan, and one could speak a little Spanish. All the women were mothers and housewives. In addition, most of the women made crafts for sale—*jippi jappa* baskets made from *Carludovica palmata* Ruiz & Pav., painted *calabashes* made from the dried *Crescentia cujete* L. fruits, bracelets, necklaces, and embroidery, though they do not have a consistent market. One woman used to make the traditional bag—the *koxtal*, but no longer does. One woman who used to make crafts now harvests crops around her house, like *coffee* (*Coffea arabica* L.) and *cacao* (*Theobroma cacao* L.). She sells the cacao to people in her own community, and the coffee to a foreign woman with a coffee shop in Punta Gorda Town. Another woman sells farm products in the community such as *peppers* (*Capsicum* spp.), *cassava* (*Manihot esculenta* Crantz), and *cocoyams* (*Colocasia esculenta* (L.) Schott or *Xanthosoma* sp.). One woman works at the *banana* (*Musa* spp.) companies or wherever she can find work during the harvest season, and makes crafts in addition. Two of the women had previously worked at the nearby ecolodge and one taught in another village for less than a year.

4.5.2 Q'eqchi' Women's Health Perspectives

A focus of these interviews was women's general perspectives on health and healthcare, as this was largely missing from my study on Q'eqchi' reproductive ethnomedicine, which focused on the specific knowledge of healers, midwives, and herbalists. To get a sense of the reproductive ailments recognized in the Belizean Q'eqchi' communities, I asked people to list the ailments that only women experience. Collectively, they listed 29 ailments or conditions specific to women (Table 4.1).

Women from the village with frequent bus access to Punta Gorda and the hospital services were asked which ailments they seek treatment for from biomedical services, which ailments they seek treatment for from traditional healers, and which ailments they treat with home remedies. Two women explained they would try the traditional healer for most common ailments, but if they could not get a successful treatment, then they would try the hospital. Similarly, hospital treatment would be sought first for something serious and if that was not successful, they would try the traditional healer next. Another woman said she would try the hospital first for everything and then try the traditional healer if the biomedical treatment was unsuccessful. One woman explained that she would go to the hospital or clinic for medicine for headaches and backaches, would see a traditional healer for heavy menstruation, and would use home remedies for children's vomiting, diarrhea, and

Table 4.1 Health ailments or conditions specific to women listed by Q'eqchi' women

Health ailments or conditions specific to women listed by Q'eqchi' women	
Menstruation	Womb pain
Menstrual pain	Womb out of place
Menstrual pain from overworking	Womb comes out of body
Heavy menstruation	Menopause and associated problems
Menstrual irregularities from falling	Stomach pain
Pregnancy	Back pain
Fever during pregnancy	Back pain from giving birth
Malpositioned fetus from falling	Back pain from frequent falling
Giving birth	Back pain from frequent pregnancy
Difficulties in birth	Back pain from falling while pregnant
Cesarean sections	Headache after pregnancy
Postpartum hemorrhage	Head and back pain from heavy work
Postpartum sickness	Headaches and weak body
Postabortion sickness	Physical problems caused by abuse
Vaginal infection	–

fever. Another woman said she would use biomedical services for high fevers, infections, sores, and "fresh cold," and would see the traditional healer for backache, womb problems and "mole drop" (fallen fontanel) in babies.

A few of the women said they would treat diarrhea, vomiting, and mild fevers in their family with home remedies, mostly medicinal plants. However, one woman said she would use biomedical services when her children suffered from vomiting and diarrhea, though she was the only woman interviewed who did not utilize home remedies or the services of the traditional healer at all. One woman said she would use biomedical services for her prenatal checkups, though another woman said she would see the traditional healer specifically for her prenatal checkups. Heavy menstruation was another reason given for seeking the treatment of a traditional healer. Diabetes is self-treated at home by one woman. *Susto*, or "fright," a spiritual disease was mentioned as another ailment treated at home.

Two women who were interviewed showed an interest in studying to be traditional healers. One woman explained that she wanted to know about plants and wanted to help people. She felt that due to discrimination from medical authorities, women healers and midwives have had be secretive of their practices, and not teach others. The other woman interested in studying traditional medicine explained that she wanted to help sick people and wanted to help other women.

4.5.3 Q'eqchi' Choices and Perceptions of Pregnancy and Birth

Q'eqchi' women were interviewed about their beliefs and decisions they made regarding their pregnancies and births. Women in the village felt that they should not do a lot of heavy work in the home, such as lift things during their pregnancies.

Some women stated that they had to do the heavy work regardless during their pregnancies, including "back" (carry) water. One woman felt that both mother and father should not have "sweets" (candy and sweet food) at night while the woman was pregnant. Another felt that pregnant women should not drink cold water or eat green pepper, but she did when she was pregnant anyway. Women stated that they should not take contraceptives, consume alcohol, or take medications during their pregnancies.

Out of the ten women interviewed in the village, four women had some of their births at the hospital and some at home, three women had all their births at home, and three women had all their births at the hospital. One of those was because the birth was by Cesarean. Of the women giving birth at home, they were assisted by the baby's father most of the time, and/or the mother or grandmother. One woman explained that the children's father delivered the babies and the mother-in-law cared for the newborns. The mother-in-law and father-in-law had explained to the father what to do during the birth and what problems he might encounter. One woman had all her births at home with a traditional healer. Another woman had half at home with her husband who is a traditional healer and the other half at the hospital with her husband present. She had all her prenatal checkups at the hospital. Only one woman said medicinal plants were used during or after the birth, and in her case were to stop the heavy bleeding after birth.

After a baby is born, Q'eqchi' women follow a number of traditional postpartum practices for mom and baby. During the first week the mother is to remain indoors, breastfeeding the baby. For 3 days they are in a curtained-off area with a continuous little fire. Cold air is believed to cause womb problems, womb pain, and back pain. Therefore, women rub a large warmed stone over their abdomens, as often as every 20 min, off and on, day and night to warm their wombs. One woman said she warmed her womb with a warmed rock every day for 1 month postpartum. The woman should be given *matz'* (porridge) with *black pepper* (*Piper nigrum* L.) to warm her and warm her womb. Also chicken soup with a lot of *garlic* (*Allium sativum* L.) so that she has resistance to illnesses. A warm cloth should be used to warm the baby especially on the umbilical cord. This is done for the first few days or as long as the first month. One mother said the baby should be given *black pepper* the first day as well.

There are various Q'eqchi' traditions practiced to protect newborn babies from evil and spirits. In some families, someone will stay awake at all times for 3 days after a baby is born and keep all the lights on through the night as well to keep spirits away. Some families will just keep the lights on for three full days to protect the baby from evil. People will pray to keep bad spirits from affecting the family. *Garlic* plays a big role in guarding from evil, and in the case of babies, it will be placed in their clothing or blankets. One mother said she would typically pin a clove on her baby's blanket. Another mother explained that when she had young children, every week, typically every Saturday, she would burn *copal* (*Protium copal* (Schltdl. & Cham.) Engl.; Fig. 4.4) and pray that nothing would harm the baby. Others said they would just pray to God or invite the pastor over to pray at the house to protect their babies.

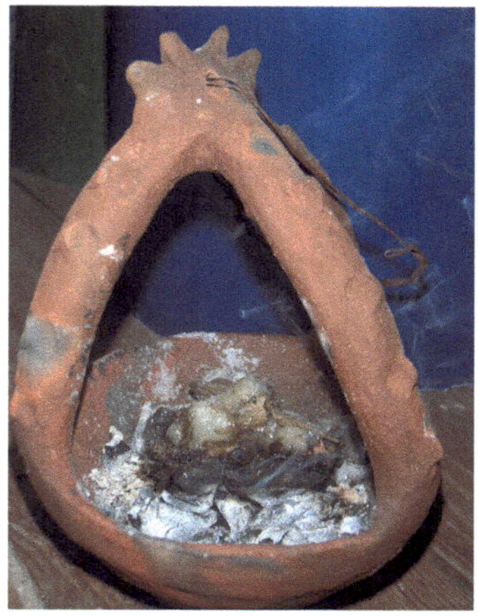

Fig. 4.4 Traditional clay incense burner with *copal* resin incense burning

The mothers interviewed breastfed their babies between 1 and 4 years each, with an average of 2 years. Most commonly used for weaning are the bitter leaves of *jackass bitters* (*Neurolaena lobata* (L.) Cass.) rubbed on the breasts to deter children from wanting to nurse. One mother said she did not do anything to wean the children, she just started giving them food. Corn *matz'* or porridge is one of the first foods given to babies after weaning. Another woman said her children breastfed for 3 years each until they could not get any more milk. During the interview her 2-year-old daughter walked up and lifted her mother's shirt to breastfeed (see mother and daughter in Fig. 4.5). This is common in Belize. Public breastfeeding, including of children who are 2–4 years old, is not uncommon.

A few beliefs regarding babies were brought up during time spent in the Q'eqchi' villages in Toledo. One, that you cannot comb a baby's hair until their teeth come in. If you comb their hair before this, their teeth will grow in spaced apart like the teeth of a comb. You are not supposed to kiss babies behind their neck because they will want to bite other children. A red thread will often be tied around the wrist to protect babies from evil and protect them from pregnant women, who are considered "hot."

4.5.4 Q'eqchi' Women's Traditions

Q'eqchi' women shared their ideas of what it meant to be a *traditional* Q'eqchi' woman. Women most commonly mentioned wearing traditional clothing as a characteristic of a traditional Q'eqchi' woman. One woman explained that appearance is indicative of tradition because from far you can see that she—in her traditional dress—is a Q'eqchi' woman. Cooking traditional Q'eqchi' food was the next most common answer followed by speaking the Q'eqchi' Mayan language. Traditional was defined by one woman as "clothing and food like our mothers'." One woman also mentioned maintaining the practice of making and consuming traditional drinks like *cacao* (*Theobroma cacao* L.). One woman stated, "I am proud to be a Q'eqchi'

Fig. 4.5 Ms. Petrona with her 2-year-old daughter who was still breastfeeding

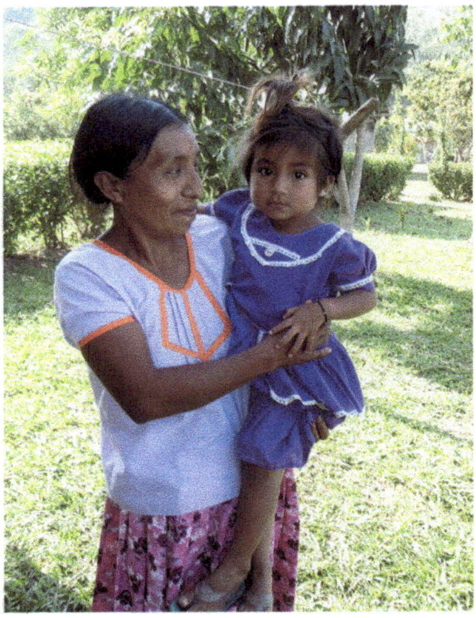

Fig. 4.6 Young Maya girls participate in a traditional dance and dress competition

woman and I maintain my Q'eqchi' identity." Another stated, "My parents were Q'eqchi' so I feel proud myself to be a Q'eqchi' woman." Another woman covered her mouth and smiled as she explained her opinion, "Q'eqchi' women were born as Q'eqchi' so they need to continue to live the Q'eqchi' way of life." One woman explained, "When I was with my parents I wore traditional clothing and kept all the traditions. When I got married my husband would buy the clothes he desired me to wear." At the annual "Maya Day" celebrated in Toledo, Maya girls can participate in a traditional dance and dress competition (Fig. 4.6).

An elder herbalist explained:

> As a traditional Q'eqchi' woman I use all my plants I have around for treating sickness in the family. I do not adapt to other people's customs. I practice a lot of Q'eqchi' customs like cutting the hair when someone is sick. Cutting the hair of the person who caused the "fright" and the person who was frightened, and burning the hair under the person sick

> from "fright." Traditions like this are being lost the quickest because people think it is a superstition. For hair and nails cut generally there is a belief to save it, so that when you die you don't search for it. Many people bury their hair so that it can't be used, or they put it between the wall boards of the house and bury it later.

This herbalist had her hair that had fallen out between the boards of her kitchen and pointed out the hair to us. When a child is sick, they cut the hair of others in the household and burn it under the child who is held over the smoke. This is done from all the other members of the household since you do not know who may have inadvertently frightened the child. This is something that traditional Q'eqchi' women do, according to two other women as well.

When the women were asked which traditions are being lost the quickest, the most common response was also traditional clothing. This was followed by traditional dances and traditional music. Other women mentioned costumes, food, and language. Plant use, including ceremonial activities such as during planting times, and cultural activities related to plants were also mentioned as traditions that are being lost most quickly in the Q'eqchi' community in Belize. Two women said "everything," in other words, all aspects of traditional Maya culture and followed up with a list of traditions: cultural dances, costumes, music, and clothing. One woman explained that "in modern days, many things have changed, things from other cultural traditions are being incorporated." A different perspective noted that "older folks have died who kept the traditions alive, like dances and cultural activities related to plants." Another agreed that "there were some customs practiced before, but due to the mixture of cultures, mostly the younger people are involved themselves in music and styles from other cultures." According to one woman, traditional clothing is getting more and more expensive. One woman noted that "many things are being lost in Q'eqchi' culture like the dances, ceremonial activities such as when it is time to plant, and traditional clothing. These are only in remote areas now." Another mother felt that "children are losing the language and the clothing because of education and their pride gets in the way."

In the past, Q'eqchi' parents would find husbands for their daughters, and it was the father's choice, some of the women explained. This still happens in remote villages, according to one woman. Her marriage was arranged. It was very difficult for her because she had to get used to him and try to love him. She feels that when you pick your own spouse it is more romantic. She would see her future husband around but did not know him and was not attracted to him. They have been together for 8 years now and recently got married. Arranged marriages are far less frequent today, but some do still occur in the Belizean Q'eqchi' villages.

Q'eqchi' women hope that their daughters will receive better educations than they did themselves, want them to be able to read, write, and speak English, and want their daughters to have more employment opportunities—not just making crafts. Women feel that education and studying hard in school will be what will benefit the futures of their daughters and granddaughters. One woman explains of herself and her husband, "I hope with the sacrifices we have made for our children they are able to study more so they can earn their living for an easier life." Another mother explained "I want my daughters to have better lives in the future. With my

Fig. 4.7 Fajina Craft Center, selling Maya women's crafts in Punta Gorda Town

15 year old [who is not in school], I hope she finds herself a gentleman who will take care of her, work peacefully, and live a good family life. Because I did not have the chance to go to school, I hope my girl in school will continue schooling for a better life in the future. I want to help them to help me and help themselves." Another mother tells that she hopes her "daughter will go to school, have the opportunity to go to UB [University of Belize], have a good life. Maybe send her to the States."

There are not many employment options available for Q'eqchi' women. Most women say that making crafts, such as baskets or weaving *koxtals*, is their only option. The women often added that they needed a better market for the crafts they were making. There is one craft shop, the Fajina Craft Center, in Punta Gorda that sells a limited amount of crafts from the women of Toledo villages (Fig. 4.7). Some women felt other possible options included working for companies like the banana companies though this involves heavy work, housekeeping or cooking, or opportunities through a women's group. My male interpreter added corn mill operation and poultry rearing. Other women simply said they could not think of any options available to Q'eqchi' women as a job or career. Women felt the reasons why it was so difficult for them to have a job or career included being busy taking care of the children in the home and for some not being able to read and write or speak English, the national language of Belize. My male interpreter added that not having the necessary experience for the position is the biggest problem Q'eqchi' women face in finding employment.

Being indigenous women in Belize, Q'eqchi' women felt most commonly that their biggest struggle was taking care of many children in the home. One woman clearly stated that, "the biggest problem is with the children. Women cannot go out because there is no one to do the housework." Another woman explained another aspect, "in Q'eqchi' tradition, it is hard for women to go out because the men get jealous, and with more than five or six *pickni* (children) it's hard to go out." Other problems women listed included their husbands being unemployed, and therefore, farming so women need to be home to prepare food for them, discrimination from other ethnic groups in finding employment, or not having a trade or training in anything in order to seek out employment. It is "hard to find jobs and money. Women have a harder time," one woman explained. Another woman stated, "I feel the shyness of Q'eqchi' women has a big effect on them in terms of seeking employment. Other ethnic groups easily cope with the environment. Q'eqchi' women are afraid to go out and seek employment and that is why they depend on men to earn all the money." Only one woman said that she does not see any problems Q'eqchi' women face in Belize.

An elder herbalist explained that childrearing is difficult for women. She has six living children and others who have passed. Her husband comes and goes, so she is basically a single parent. To make money, she has a small piece of farmland that she grows things on, a 1 h walk from her house. She grows okra, calaloo, sweet pepper, cabbage, pumpkins, cucumbers, papayas, limes, culantro, and any other seeds of food plants she finds. She sells at the market once or twice a week in Punta Gorda Town. She brings her products to the market and also walks around selling door to door and to people on the street. She is one of a small percentage of women who grow all the vegetables they sell at the market. Q'eqchi' individuals who have significant knowledge about medicinal plants, both male healers and female herbalists, seem to be growing a larger diversity of food plants than the majority of people in their villages.

4.5.5 Gender Differences in Healing and Plant Use

Traditional knowledge and practices are often gendered, with girls learning from women and boys learning from men. E. N. Anderson (2003) writes that even in small Yucatec Maya communities, the pool of knowledge is dispersed, with no one individual knowing the full body of traditional medical knowledge or plant names. Anderson explains that:

> In Maya society, men do most of the forest-based activities (hunting, chicle gathering, logging etc.) and know tree names and forest-plant names better than women. Men and women are both active in milpa and orchard, and are equally knowledgeable about plants of those environments... Gender also structures nomenclatural knowledge. Women manage most of the home medicine and health care, and thus know more medicinal plant names than men (other than male herbalists). Women also are in charge of the ornamental and medicinal section of the garden, and thus are the ones who know most of such plants.

In the Belizean Q'eqchi' communities, women typically grow plants around the house and would collect plants from the community, roadsides, and the edges of forests. Men are more likely to go deep into the forest to look for plants. There seemed to be some differences in the species used between men and women due to their different collecting sites. This is also tied into the largely gender-specific lines of knowledge transmission in Q'eqchi' Maya culture. However, today many Q'eqchi' women have learned from their husbands or fathers, as women healers are very few in number. One woman says regarding the plants she uses that she "grows some here at the house, some are common[ly found], and some my husband has to find in the bush." One woman says she gets her plants around her house near Punta Gorda, and some from out in a more remote village.

4.5.6 Q'eqchi' Midwives and Healers in Belize

In 1981, Kathryn Staiano wrote that in Belize most rural midwives were able to practice without a license, although there were some government attempts to increase their training. The rural midwives offered prenatal care as well as delivered babies. She reported that even in Punta Gorda Town, the unlicensed midwives gave prenatal care in the home, but that pregnant women living in town delivered their babies most often in the local hospital.

Since I began research in Toledo in 2007, I have only heard of four Q'eqchi' midwives, and one passed away before I had the chance to meet her. Three Q'eqchi' midwives were interviewed in Belize, and only two were still practicing. One had moved to Belize from Guatemala just 5 years before and was unsure if she was legally allowed to train other women in Belize who were interested in learning midwifery and was afraid it could be "used against her." The other midwife was "too shy" or afraid to have the interview with me and arrived very early to the interview so that she could interview only with the Q'eqchi' woman who was my interpreter and research assistant.

The midwife who was no longer practicing midwifery now considers herself a "woman healer." She learned what she knows from her husband who was a traditional healer, although she has been widowed now for some time. She stopped assisting during deliveries a decade ago because of her eyesight, but she had delivered babies for 12 years. Now she does massage and positioning the baby in the womb. She sends them to the hospital for the birth, then they bring the baby back and she massages and warms them. She can also turn breech babies in the womb, using external cephalic version. She explained that in her current practice, transportation around the district to see her patients can be a barrier. She says some people will come to see her for treatment, others will come pick her up, and sometimes she is then responsible for her own transportation for follow-up visits. Sometimes when there is no transportation for her, she has to give up the work because it does not pay enough.

The midwife who had moved to Belize from Guatemala 5 years before, shared some of her personal history and training. She came out of school in 2nd grade, and was trained in midwifery in a clinic in Guatemala, when she was around 30 years old. Five hundred Q'eqchi' women learned midwifery there and were taught by Mexican midwives. It was a 3 year training. She had delivered around 40 babies in Guatemala and about 7 in Belize, in two different villages. The first woman she helped in Belize was her daughter-in-law, and word of mouth spread from that birth. She is the only midwife in her village. She does massage during pregnancies, delivers babies, treats menstrual problems, "mole drop" (fallen fontanel) in babies, babies' and kids' ailments, and teenagers with menstrual problems.

No other midwives were located. It is possible that some are practicing secretly, or that my networks did not reach far enough to hear of them all. However, most of the people I considered knowledgeable enough to ask their advice on where to find midwives, said there were none at all, nor any women healers that they knew of.

Women with medicinal plant knowledge in the Belizean Q'eqchi' community I refer to as herbalists, but they are simply known among their circle of family, friends and neighbors, to be women with this sort of knowledge. Some are quite knowledgeable, with levels perhaps equal to that of the male *ilonels* or traditional healers, whereas other women have basic home remedy knowledge for treating common ailments that they have learned from their family members and through their own interest. One elder herbalist knows a bit about traditional Maya acupuncture, a practice very few people know about anymore.

4.5.7 Loss of Female Healers in the Belizean Q'eqchi Community

In many parts of Latin America, traditional birth practices and the role of the traditional midwife are being changed and influenced by the spread of biomedicine, through midwifery and T.B.A. training programs, through radio announcements and many other ways (Cosminsky 1994). Michel et al. (2007) published on medicinal plants used by the Q'eqchi' in Livingston, Guatemala, just an hour by boat across the Bay of Honduras from Punta Gorda, Belize. Both Q'eqchi' people, in Guatemala and in Belize, live primarily farther inland from the coastal town centers of Punta Gorda and Livingston. Michel interviewed practicing Q'eqchi' midwives, called *comadronas* in Spanish as well as male Q'eqchi' traditional healers, called *curanderos* in Spanish, in her work.

There are very few practicing midwives in the Q'eqchi' communities in Belize, however, just an hour across the Bay of Honduras, in Livingston Q'eqchi' midwives traditionally assist in births, as well as treat all complaints related to women's reproductive health. In the Belizean Q'eqchi' community, women who are in remote villages without easy or affordable access to the hospital services, are mostly giving birth at home with only the help of their husbands or sometimes other family members. If difficulties arise, sometimes a male traditional healer is called to help. Male traditional healers have begun to treat women's reproductive health ailments as well in the Belizean Q'eqchi' communities because there are no longer practicing female traditional healers or midwives. This is neither traditional nor seen as culturally appropriate, by anyone in the community, but is practiced in this way out of current necessity.

An elder Q'eqchi' woman who is a midwife, but no longer practicing explained that "there were women healers in the past, and they specialized in women's and baby's health." She thought that there were probably still more in Guatemala. The question is why are there no longer practicing female traditional healers or midwives in the Belizean Q'eqchi' community, especially when there are still Q'eqchi' midwives practicing in Guatemala (Michel et al. 2007). There seem to be a number of factors involved. The retired midwife felt that there are no longer women healers in the Belizean Q'eqchi' community because of lack of interest and loss of tradition. A male traditional healer agreed that young people are not interested and the loss of interest is due to modernization in the Belizean Q'eqchi' community. Mopan communities of southern Belize have also had a decline in traditional healing "because the younger generation lacks interest in it and because evangelical missionaries discourage it" according to Steinberg (1997, 1998).

Still, other Q'eqchi' young women say they are in fact interested, but unable to pursue those interests, because of more practical reasons. They say they are interested in learning traditional medicine, and apprenticing with an elder healer, but do not have the money to pay for travel, training, and childcare. There are so few women who are known as traditional healers or midwives (practicing or nonpracticing) that the young women wishing to learn would need to travel to another village, which is not feasible for women with their domestic and childcare responsibilities in their households. One woman explained some of the issues for her and other Q'eqchi' women, stating it was hard to go out to apprentice and later to treat patients because husbands get jealous, and hard to go out when women have several children, which is common.

Male healers have started to treat ailments that were traditionally the realm of female healers and midwives, such as women's reproductive ailments, as well as sometimes delivering babies, in place of a midwife. A particularly revered male healer explains he specializes in women's health because so many people seek him out for these ailments. Another male healer said that "there are no ladies to treat these sorts of things, so the male healers will only when absolutely necessary. In the past, when I was growing up, there were women who dealt with those problems in particular. We know jealousy exists. Treating those types of ailments [women's reproductive health] can bring problems because I am a man and am not sure what the patient's husband might be thinking." He has only delivered his wife's babies.

Another male traditional healer says he would only deliver a baby if a member of the family was unable to, and he had their consent. When he was living in one of the remote villages, there was no hospital or clinic so he would go and check the position of the baby and massage the woman and help to speed delivery, then would leave it in their hands if he could. He would return to do a postpartum check and "heal the bones and set them back into place." The Q'eqchi' midwife that had moved to Belize just 5 years before from Guatemala wants to train Belizean women, who have expressed interest in learning midwifery, but is unsure if this is "allowed" in Belize and is afraid that it could be "used against" her. One mother had the range of experiences—one child at home with a midwife, one child at home without a midwife, and one child at the hospital. She said the birth with the midwife was the best experience. She used "bush medicine" (medicinal plants) and had the baby quickly.

Although there are few practicing midwives, there are women knowledgeable about medicinal plants, not considered healers who I refer to as herbalists. Who have these women with medicinal plant knowledge learned from then? An elder midwife who no longer does deliveries, only prenatal and postpartum care says she learned to treat pregnant women and women and babies after birth from her husband, who was a healer. He had the knowledge, but the community did not know, so he trained her to do the work because he felt it was better and more appropriate for her to do the treatments, but he would usually accompany her, she explained. Another elder herbalist learned from her husband and father. The wife of a traditional healer also explained that she learned what she knows from her husband. Some female herbalists learned from their fathers, uncles, or other family members who were *ilonels* or traditional healers, some learned from their mothers who were also herbalists. A couple of the herbalists had sisters who became midwives, so the knowledge and interest ran in the family. Two women said they had learned some from the bush doctors who have treated them. Women said they learned from family, in general, parents, mothers, grandparents, uncles, or elders, friends, and neighbors. The male *ilonels* all learned from another male family member who was an *ilonel*, except for one who was sent by his father to learn from a nonrelated male *ilonel*.

Domestic obligations and cultural limitations to women's far and long-term travel keep women bound to their homes for the most part. According to a revered male healer, "women used to learn from mother and grandmother, but if they don't live with or near them they would need to travel out of the house to go to someone's house to learn about healing, typically after dark, after the day's work is done when there is 'more power'." He further explained that, "it used to be that women healers treated reproductive health ailments, but not many women learn it anymore because you learn these things at night and women are afraid. There is more power at night that is why they learn at night. Women used to learn from their mother, grandmother and other family members, but now they would have to go out at night to someone else's house." It is not very culturally accepted for women to be out at night, so this may be a part of the fear that women have, that this healer is referring to.

Regarding Q'eqchi' women's mobility, Warner (2007), who conducted research on Q'eqchi' women from Guatemala, living in a refugee camp in Maya Tecún, Southern Mexico, noted that married Q'eqchi' women are traditionally permitted to visit their mothers on a regular basis. Q'eqchi' women's mobility is culturally limited otherwise, so this visitation right allows her some mobility and keeps her from being isolated. Warner explains that:

> In Maya Tecún, each time a woman leaves her home, she takes the risk of coming into contact with a person or object that will affect her balance of hot (q'ix) and cold (ke), thereby endangering her fertility and the health of her children. A complex system of social control through fear of retribution by the Tzuultaq'a (mountain spirits) traditionally provides the ideological basis for restricting women's movements and contact with non-kin members of the community.

Q'eqchi' women often responded *ink'a' ninnaw* (I do not know) shyly to my questions during interviews. Often they did know, and with a little patience and persistence on our (my and my interpreter's) part, the women would share more details of what they knew. Lack of self-confidence (and experience, or self-confidence to gain the experience) is another factor limiting to Belizean Q'eqchi' women in pursuing studies of traditional medicine. A woman in her early twenties explained that she only treats her family members with medicinal plants, and said while giggling "because I am young and don't have experience and don't want to kill people." Many women only treat within the family. A couple of women explained that if someone not related to them asks for treatment, the women would just give instructions and have the person seeking treatment, treat themselves as per their instructions. They were both knowledgeable about medicinal plants, but like others, lacked the confidence to practice what they knew more formally.

Some women show a bit more confidence and see the importance of their knowledge. A female herbalist shared, "I hope my daughters will learn traditional ways of the bush like bush medicine, and also learn and do well in school." Another herbalist says that she sometimes goes to the "bush doctor" (traditional healer), however, usually she tries first herself to treat the ailment. Although she does not consider herself a healer, she is very knowledgeable and her father is a bush doctor and sister is a midwife, both in Guatemala. This woman will first try to treat things herself, then try a local bush doctor if it does not work, and then the hospital as the third option.

According to some, women healers experienced discrimination from medical authorities so chose to be secretive about their practice. One woman says she is afraid to "let on" publicly that she knows so much about medicinal plants. She does not work like her father does, who is an *ilonel* (traditional healer) in Guatemala, she just treats within the family. Her sister is a midwife also in Guatemala. She says there are more female healers practicing still in Guatemala. According to one *ilonel* in Belize, there are a few women who know about traditional healing and medicinal plants in the Belizean Q'eqchi' community, who just need to be encouraged, like his stepsister. At the beginning of an interview with her she said that she does not use plants, just massage. During the course of the interview, however, she mentioned several plants that she knows about and uses.

Secrecy may be a factor in determining accurate numbers of midwives and female traditional healers practicing in Belize. Three Q'eqchi' midwives were interviewed in Belize during this study. One elder Q'eqchi' woman was "too shy" to do the interview with me and interviewed only with my female Q'eqchi' research assistant, another midwife says she is no longer delivering babies and just does prenatal and postpartum visits, and the third midwife who had moved to Belize from Guatemala 5 years earlier said she had women in Belize ask her to teach them, but she said no since she was afraid because she was unsure of the laws. Admittedly, the legalities of the matter do seem purposefully ambiguous since there are differing opinions as to what is legal. Does this indicate that there may be more midwives who do not want to admit that they deliver babies? Even though untrained husbands admit to delivering their wives' babies in their homes?

Female healers, who traditionally will often also practice midwifery, may have more reasons for secrecy of their traditional medical practice than the male *ilonels* who do not traditionally assist during childbirth. However, in some cases in this research, men have been secretive about what they know regarding traditional medical knowledge. Two male healers said they knew a plant for every ailment I mentioned but said they did not know the name and would often describe the preparation or administration, but they did not share any plant names. I also interviewed the wife of one of these traditional healers, however, who is a midwife, and she shared much more with me than her husband had.

4.5.8 Q'eqchi' Migration

A look at Q'eqchi' migration from Guatemala to Belize is critical to understand the absence of many female healers in Belize. The Q'eqchi' are originally from Alta Verapaz, Guatemala, and have been displaced numerous times, migrating to escape landlessness and debt peonage (Warner 2007). Wilk (1997) discusses the history of Q'eqchi' movements, noting that:

> The terms *dependence* and *flight* encapsulate a dilemma that has been and remains at the core of the Kekchi experience. From the time of the conquest, loss of political independence, land, control of labor, and even the most basic human rights have forced many Kekchi into the position of a dependent peasantry or proletariat...History shows a clear relationship between the weight of oppression on the Kekchi and the number who choose flight.

Wilk (1997) discusses how migration to remote areas was not voluntary and that most Q'eqchi' would have preferred to continue living independently in their fertile homelands, where there were trade and markets, as well as Maya social and cultural institutions. Unfortunately, this was not an option for most. He goes on to explain that:

> Isolated farms in unfamiliar climate, far from medical care, economic opportunity, or relatives, are a last resort for those fleeing something even worse...Many who fled into the lowlands have continued to move, in search of that ideal combination of access to land, freedom from oppression, and access to markets, roads, schools, and other amenities. Although the Kekchi in Toledo District appear to be forest Indians emerging from a primeval economy into a developed one, they are actually survivors of an exodus, emerging from the waters into what they hope is a promised land.

Many older Q'eqchi' people came to Belize 50–60 years ago to escape hard forced labor in Guatemala. Two of the male healers' parents had brought them to Belize when they were young. One healer's parents brought him to Belize when he was a young boy because they had been forced into hard unpaid labor like breaking rocks and other difficult work for railroad construction. In Belize, his parents made their own *milpa* and grew food and livestock for the family. According to one healer whose family brought him to Belize when he was a young teenager, his parents had been forced to do hard labor for 1 or 2 cents per day in Guatemala. His parents came to Belize and worked in the banana plantation. One of the eldest male healers came to Belize because he himself was a forced laborer in Guatemala, working 5 years with no pay. He said whenever he had 10 or 25 cents that was a lot. He was tired of it, and came to Belize approximately 50 years ago. Both of his parents had been forced to work harvesting coffee for no pay as well in Guatemala.

Some Q'eqchi' came around 25 years ago from Guatemala seeking medical treatment or other services offered in Belize. One traditional healer came around 25 years ago seeking a better education for his children. He was living just across the border from Belize in Guatemala. He was a member of a cooperative to see how best they could provide their children with a good life. The other side of the border in Belize had a school in Jalacte, so he moved there for his children's schooling. Then, a decade later he moved his family to the Indianville neighborhood of Punta

Gorda Town, again for his children's education, and for work opportunities. Some men and women came to Belize from Guatemala in more recent years, in the last 10–15 years primarily seeking farmland, or seeking employment for a better life.

There is also movement of Q'eqchi' people within the district of Toledo as well, as the continuation of the migratory movements from Guatemala. Wilk (1997) explained that Q'eqchi' migration is not solely linear movement from Guatemala to Belize. He reported that in 1978 adults older than 20 years of age within the Toledo District had moved an average of every 9.7 years in response to changing economic opportunities, land availability, local politics, and familial relationships.

Traditionally, the bond between mothers and daughters is the strongest social relationship that Q'eqchi' women have. Q'eqchi' women from Guatemala living in a refugee community in Mexico identified their most important source of social support at all stages of life were their mothers, while it was recognized that they rarely received support from their neighbors, friends, *compadres*, or those they were related to through marriage. Q'eqchi' women whose mothers were not present in the camp displayed an increase in stress-related symptoms, including nervousness, headaches, and stomach ailments. Traditionally, before displacement, Q'eqchi' families arranged and encouraged marriages between families located near one another, and mothers were able to maintain active roles in their daughters' lives (Warner 2007). Warner (2007) further explains:

> It was unusual for women to visit women who were not related to them, in that a woman's social relations were traditionally very much limited to her kinship group. Typically, there was a significant support extended to women from their natal families in the form of food, information, assistance in daily tasks, and emotional support.

Warner (2007) goes on to assert that Q'eqchi' women who do not live near their mothers or have weak natal kin support experience ruptures in the transmission of traditional gendered knowledge related to healing, rituals, food preparation, and caring for children. Traditionally, Maya women would have learned about these aspects of life from their mothers. Furthermore, "lack of natal kin isolates and silences Q'eqchi' women in that they are restricted to their homes and have no means of communication with other women."

According to Joanna Michel's 2006 study in Guatemala, "most Q'eqchi villages have at least one midwife who specializes mostly in pregnancy and childbirth, but also treats other gynecological conditions including menstruation, infections, and menopausal symptoms." It is possible that women healers, which are quite often also midwives, migrate less frequently because of strong ties to the community they live within. I interviewed an elder Belizean Creole midwife in the Cayo District, who said she had delivered nearly every person within a large generational range in her town, when everyone birthed at home with a midwife. She said that the children being born to many of those people now are being born in the hospital, but she can look at a group of adult Belizeans in town and say she had delivered each one. She still gets called for births. She is loved by everyone, and known by everyone. She is not a Q'eqchi' woman and lives in one of the major towns of Belize, but her story is illustrative of the bonds that midwives often develop with so many individuals as well as the community on a whole. This may in turn mean midwives are less likely to migrate, compared to women without this tie, and without an occupation. Many male healers supplement their income with farming and wage labor, and may be more likely to migrate. One elder female herbalist told me I was "too brave" to come so far to Belize by myself.

Conclusions

Women's traditional knowledge in the Q'eqchi' communities of Belize is disproportionately fragmented. A retired midwife explained that women's knowledge and knowledge of treatments for women's health are the first to be lost. She says in the past there were more women healers, and they specialized in women's health, but there are not really any more in Belize due to lack of tradition and lack of interest. She says she thinks there are probably more women healers still in Guatemala. When traditional practices begin to disappear this is an example, likely not isolated, in which women's knowledge appears to be disappearing at a faster rate. Men's knowledge is also becoming fragmented, but less so, partially because many of their networks have stayed farther-reaching since they have not been tied to domestic obligations to the same extent as women.

Many of the women who know a lot about medicinal plants, also have husbands who are recognized "bush doctors" and often say they have learned from their husbands, although sometimes also from a father, uncle, or other male family member. However, these same women always know additional remedies and use additional medicinal plants to what their husbands use. So it seems women are attributing more to their husbands or male family members, than is really due. Male traditional healing knowledge has become more greatly valued and, therefore, seen as the source of information that one learns. This has perpetuated the situation in which women are no longer seen as traditional healers in the Belizean Q'eqchi' community, even though they may very well be healing, with traditional plant-based medicines, as the male healers are. It seems then that women are more likely to attribute male healers as the source of their knowledge, even if they have learned from knowledgeable women in their lives, and in turn, devalue their own knowledge and abilities to practice traditional medicine and teach younger women.

Knowledge of reproductive health treatments, traditionally held by women, is disappearing more rapidly than other areas of traditional medical knowledge according to the community. This is due to multiple factors—immigration from Guatemala disrupted familial networks and, therefore, the most traditional and common means of knowledge transmission; many midwives may have stayed in Guatemala as they had occupations and their lives were strongly tied to the communities they had delivered multiple generations of babies in; practicing lay midwifery has been discouraged by the Belize MOH and, therefore, some women are afraid to practice and teach others what they know; women are not engaging in traditional training with healers or midwives because it is not feasible for them to travel to other villages or even homes to study because of domestic obligations and financial limitations;

and knowledgeable women often lack the confidence and do not consider themselves to be healers—if others devalue their knowledge, so will the women. These factors combined have led to the accelerated loss of women's healing traditions in the Q'eqchi' Maya communities of Belize.

Most traditional knowledge is inherently gendered. Indigenous Chinantec couples in Oaxaca, Mexico were found to have very little shared medicinal plant knowledge between spouses (Browner 1991). Chinantec men are also knowledgeable in medicinal plants for women's reproductive health, and Browner (1991) asserts that the men "enhance their influence over the reproductive process by becoming knowledgeable about herbal remedies used to regulate female fertility and treat reproductive health problems." Women have their own traditional beliefs, practices, remedies, and wisdom that is distinct from men's and serves the purpose of meeting women's reproductive objectives and needs as they themselves identify them. Women's intergenerational knowledge transmission is critically important for the maintenance of biological and cultural diversity (Howard 2003).

The remaining Q'eqchi' female healers and midwives in Belize are mostly in their 60s and 70s. Undoubtedly as they pass away, a great deal of knowledge is lost with each elder. Traditional Maya midwifery training programs should be provided in conjunction with the MOH's T.B.A. training programs. The T.B.A. programs have historically trained women who were already midwives or had some relevant experience. Since there are no women training to be traditional midwives in the Belizean Q'eqchi' communities, there will not be any younger women with that deeper base of knowledge needed for practicing confidently after a several month T.B.A. training. The MOH's services are not accessible to Q'eqchi' women in the remote villages where women are giving birth at home often with only the help of their husbands. Those communities need the services of traditional midwives and women's traditional knowledge needs to reestablish the value it was once attributed in Q'eqchi' Maya culture, for the health, wellbeing, and livelihoods of Q'eqchi' women in the Toledo District of Belize.

References

Anderson EN (2003) Those who bring the flowers: Maya ethnobotany in Quintana Roo, Mexico. ECOSUR, Chetumal

Browner CH (1991) Gender politics in the distribution of therapeutic herbal knowledge. Med Anthropol Q 5(2):99–132

Cosminsky S (1994) Childbirth and change: a Guatemalan study. Ethnography of fertility and birth, 2nd edn. Waveland Press, Prospect Heights

Crooks DL (1997) Biocultural factors in school achievement for Mopan children in Belize. Am Anthropol 99(3):586–601

Hinojosa SZ (2004) Authorizing tradition: vectors of contention in Highland Maya midwifery. Social Sci Med 59:637–651

Howard PL (2003) Women and the plant world: an exploration. Women & plants: gender relations in biodiversity management & conservation. Zed Books, London

Jenkins GL (1993) Burning bridges: policy, practice and destruction of midwifery in rural Costa Rica. Soc Sci Med 56:1893–1909

Kufer JK (2005) Plants used as medicine and food by the Ch'orti' Maya: ethnobotanical studies in eastern Guatemala. Dissertation, University of London

Maraesa A (2009) I no 'Fraid for that: pregnancy, risk, and development in southern Belize. Dissertation, New York University

Michel J (2006) Medical ethnobotany of the Q'eqchi Maya: perceptions and botanical treatments related to women's health. Dissertation, University of Illinois

Michel J, Duarte RE, Bolton JL, Huang Y, Caceres A, Veliz M, Soejarto DD, Mahady GB (2007) Medical potential of plants used by the Q'eqchi' Maya of Livingston, Guatemala for the treatment of women's health complaints. J Ethnopharmacol 114:92–101

Staiano KV (1981) Alternative therapeutic systems in Belize: a semiotic framework. Soc Sci Med 15B:317–332

Statistical Institute of Belize (2007) Abstract of statistics. Belmopan, Belize

Steinberg MK (1997) Death of the dance: cultural change and religious conversion among the Mopan Maya in Belize. Cult Surviv Q 21(2):16–17

Steinberg MK (1998) Mopan Maya forest resources in southern Belize. Geog Rev 88(1):131–137

Warner FR (2007) Social support and distress among Q'eqchi' refugee women in Maya Tecún, Mexico. Med Anthropol Q 21(2):193–217

Wilk RR (1997) Household ecology: economic change and domestic life among the Kekchi Maya in Belize. Northern Illinois University Press, DeKalb

Zarger RK (2002) Acquisition and transmission of subsistence knowledge by Q'eqchi' Maya in Belize. In: Stepp JR, Wyndham FS, Zarger RK (eds) Ethnobiology and biocultural diversity. International Society of Ethnobiology, University of Georgia Press, Athens

Chapter 5
Conclusions

This ethnomedical study was intended to document and preserve traditional knowledge related to reproductive health, with a focus on the plant species used. While the Q'eqchi' Maya still maintain a rich ethnopharmacopoeia of medicinal plants for reproductive health, the collective knowledge is dwindling with the passing of each generation of healers, midwives, and herbalists. The results of this collaborative study make a contribution to revitalizing that knowledge through the creation of this reference that can be a resource for training Q'eqchi' youth who want to learn traditional medicine. In conclusion, I will review the three hypotheses from my study, summarize the research results, and discuss some points of broader contextualization to consider.

5.1 Q'eqchi' Reproductive Ethnomedicine

My first hypothesis was: *Q'eqchi' traditional medicine has an extensive ethnopharmacopoeia of medicinal plant species used for reproductive health, as well as culturally specific ethnomedical practices*. I interviewed 6 male traditional healers, 3 female midwives, and 12 female herbalists in the Q'eqchi' Maya communities of Belize. There are at least 60 species used to treat reproductive health ailments among the Q'eqchi' Maya in Belize, in 50 genera, and 36 plant families. In this study, 52 species have been identified to species, and 8 to genus. While plants referred to by 41 different common names have not been collected, or identified if they were collected, some of these 41 may be conspecifics to those already collected or those referred to by other common names. Piperaceae was the plant family with the most species used in Q'eqchi' reproductive ethnomedicine, followed by Fabaceae and Malvaceae (Table 5.1).

An interesting pattern emerged, with similarities in species use down gender lines, in Q'eqchi' medicinal plant knowledge. Table 5.2 shows species used commonly, here defined by at least 25 % of females (total $N=15$) or males (total $N=6$). Note that there are plants of the same genus used commonly by both genders, from the genus *Piper*, however this is from the plant family Piperaceae, with the highest

J. De Gezelle, *Q'eqchi' Maya Reproductive Ethnomedicine*,
SpringerBriefs in Plant Science, DOI 10.1007/978-3-319-10744-8_5

Table 5.1 Plant families with the most species used in Q'eqchi' reproductive ethnomedicine

Plant family	Number of species
Piperaceae	6
Fabaceae	4
Malvaceae	4
Melastomataceae	3
Apiaceae	2
Asteraceae	2
Cucurbitaceae	2
Lamiaceae	2
Lauraceae	2
Myrtaceae	2
Poaceae	2

Table 5.2 Most common species used by Q'eqchi' female midwives, herbalists, and male healers

Most common species used by females	Number (percent) of women using species	Most common species used by males	Number (percent) of men using species
Momordica charantia	8 (53 %)	*Clidemia crenulata*	3 (50 %)
Piper nigrum	8 (53 %)	*Maranta arundinacea*	3 (50 %)
Theobroma cacao	8 (53 %)	*Piper peltatum*	3 (50 %)
Neurolaena lobata	6 (40 %)	*Piper tuerckheimii*	3 (50 %)
Allium sativum	5 (33 %)	*Desmoncus orthacanthos*	2 (33 %)
Pimenta dioica	5 (33 %)	*Psychotria acuminata*	2 (33 %)
Citrus aurantifolia	4 (27 %)	*Tococa guianensis*	2 (33 %)
Ocimum campechianum	4 (27 %)		
Syzygium aromaticum	4 (27 %)		
Zea mays	4 (27 %)		

number of species used in Q'eqchi' reproductive ethnomedicine, and is the only similarity between the two genders' commonly used plants. It is clear that while Q'eqchi' reproductive ethnomedical knowledge is the domain of both male and female healers and herbalists, it is highly gendered.

Women and men are using different species to treat reproductive health ailments; in fact none of the species most commonly used by one gender (used by at least 25 % of the individuals) are commonly used by the other gender. Female midwives and herbalists are primarily utilizing common weedy species like *Momordica charantia* and *Neurolaena lobata*; culinary spices like *Allium sativum*, *Piper nigrum*, *Pimenta dioica*, *Ocimum campechianum*, and *Syzygium aromaticum*; and food plants like *Theobroma cacao*, *Citrus aurantifolia*, and *Zea mays*. By stark contrast, all the species commonly used by the male healers are found growing in forested areas (Fig. 5.1), with only one species *Piper peltatum* growing primarily in successional or riparian areas. The species used by the male healers are not weedy, cultivated or eaten as food like the plant species commonly used by women.

Fig. 5.1 Don Francisco collecting medicinal plants in a forested area of Toledo

The most medicinal plant species in Q'eqchi' reproductive ethnomedicine were used during labor and childbirth (24 species; Table 5.3). This was followed by menstrual pain (22 species), postpartum treatments (21 species), female infertility (15 species), and menopausal symptoms (14 species).

From a biomedical perspective, some traditional Maya medical practices and concepts may be difficult to understand. Interestingly, many of these practices and concepts are found in cultures outside of the Maya world. As the biological assay results have shown, there is great wisdom behind the Maya medicine still practiced in Central America today. Corroborations by cultures on other continents add another level of validity, beyond that inherent in the maintenance of a practice or concept within a single traditional medical system. Consider these examples of culturally specific reproductive health ailments from other cultures. Regarding womb disorders, Ward (1977) discusses a disease from the island of Pohnpei, Federated States of Micronesia that is strikingly similar to *k'uub'sa'* in Q'eqchi' ethnomedicine. The Pohnpeian womb disorder is called *soumwa en pahsoan* or "foundation sickness" that can be caused similarly if a woman falls, carries heavy loads, or engages in too much strenuous activity. Like *k'uub'sa'*, foundation sickness can also result in a range of reproductive ailments such as abdominal pain, irregular menstruation, miscarriages, or infertility. He explains that:

> If the "foundation" works properly, a woman has regular periods, conceives and bears healthy children, and other internal organs also function smoothly. The trouble with the foundation organ is that it tends to move around inside the body. When a woman falls or performs too much exercise, it can be jarred out of place…A married woman whose foundation is out of place is unable to conceive, and the pains initially experienced in the uterine area may spread to other parts of the body (back, legs, joints, and so on).

Commonalities of postpartum precautions across medical systems exist as well. Women in Asia and in Latin America will each follow a period of physical confinement after giving birth often in a warmed area, and follow other practices and restrictions in order to protect her at this vulnerable time from the cold and wind, and maintain her humoral balance (Manderson 1981). Variations on these practices have been found among Malay women (Manderson 1981) and rural Korean women (Sich 1981). These periods of confinement help to ensure the new mother rests and avoids contact with carriers of infections (Manderson 1981).

Table 5.3 Number of species used for categories of reproductive health conditions

Condition category	Number of species
Labor and cildbirth	24
Menstrual pain	22
Postpartum treatment	21
Female infertility	15
Menopausal symptoms	14
Female contraceptive	12
Miscarriage prevention or treatment	10
Amenorrhea (lack of menstruation)	9
Galactagogue (increases breast milk)	9
Hemorrhage (uterine and postpartum)	9
Uterine fibroids	7
Heavy menstruation	7
Womb disorders (*K'uub'sa'* and "Cold Inside")	7
Delayed menstruation	6
Placenta expulsion	6
Abortifacient	5
Pregnancy	5
Body pain from pregnancy and birth	4
Male impotence and infertility	4
Morning sickness	4
Infection (vaginal or uterine)	3
Irregular menstruation	3
Male contraceptive	1
Pregnancy test	1
Swollen testicles from *Obeah*	1

Recognition, prevention, and treatment of womb disorders including medicinal plants, abdominal massages, postpartum steam baths, postpartum abdominal warming with a heated rock, and postpartum banding of the abdomen constitute a significant area of Q'eqchi' reproductive ethnomedicine devoted entirely to uterine care. Perhaps biomedicine has lessons to learn from Maya and other traditional medical systems on this topic. Prolapsed uterus may correspond or overlap to some degree with the womb disorders recognized in Q'eqchi' ethnomedicine. This may be underdiagnosed in biomedical practices or not considered the cause of other conditions like heavy menstruation and infertility, as with Q'eqchi' womb disorders.

According to the Centers for Disease Control and Prevention (CDC 2013), in the USA, Cesarean sections and hysterectomies are the two most frequently performed major surgical operations for women of reproductive age. Approximately 600,000 hysterectomies are performed in the USA every year. Between the years 2000 and 2004, the three conditions most often associated with hysterectomy were uterine leiomyoma ("fibroid tumors"), endometriosis, and uterine prolapse (Whiteman et al. 2008). Uterine fibroids are noncancerous growths on the uterus that can cause bleeding, pain, infertility, and pregnancy complications. Uterine fibroid proliferation is estrogen-dependent, and selective estrogen receptor modulators (SERMs) have been shown to inhibit growth in fibroid cells (Fuchs-Young et al. 1996). Endometriosis is a complex condition characterized by cells from the lining of the uterus (endometrium) growing outside of the uterus, which can lead to pain, bleeding, and infertility. Endometriosis is also estrogen-dependent, and current therapies focus on lowering endogenous estrogen levels (Rizner 2009). Uterine prolapse is the falling or sliding of the uterus from its normal position, and has actually been associated with low levels of endogenous estrogen in premenopausal women (Lang et al. 2003).

Q'eqchi' Maya reproductive ethnomedicine emphasizes womb health, thermal equilibrium, hormonal equilibrium, and healthy fertility regulation. Given the frequency of such conditions as endometriosis, uterine fibroids, infertility, and reproductive cancers in the world, worthwhile lessons may be gleaned from the Q'eqchi' ethnomedical conception of maintaining reproductive health. Perhaps uterine massage and estrogenic plant-based medicines can provide an alternative to hysterectomy in some cases. The position of wombs is regularly considered in Q'eqchi' ethnomedical diagnoses related to reproductive health, and care is taken to maintain womb health starting from menarche. If malpositioned wombs and the hormone imbalances are diagnosed and treated early as they are in Q'eqchi' traditional medicine, perhaps hysterectomies and other surgical procedures will not be needed in some of the biomedical cases.

5.2 Estrogenic Plant Use in Q'eqchi' Reproductive Ethnomedicine

My second hypothesis was: *Traditional Q'eqchi' medicine treats conditions related to hormone imbalances with plants containing compounds whose bioactivity can be demonstrated in standard hormone bioassays*. Out of the ten species selected for bioassays, nine displayed novel estrogenic activity, four also displayed novel antiestrogenic activity, and two of the extracts were cytotoxic to the breast cancer cells used in the assay. Bioactivity guided fractionation and isolation of estrogenic constituents would further deepen our understanding of the efficacy and mechanism of action of medicinal plants used in Q'eqchi' reproductive ethnomedicine.

Four of the species, the three from the Melastomataceae family, as well as *Psychotria acuminata* have morphological characteristics that correspond with their uses. The Doctrine of Signatures refers to human selection of medicinal plants based on some physical characteristic of the plant that resembles a human body part or human body characteristic, with the belief that these plant characteristics are suggestive of the plant's medicinal application. For instance if the plant has a structure that resembles a human body part, the plant would be used medicinally to treat ailments of that body part. Scientists and skeptics often assume that plants used medicinally based on the Doctrine of Signatures must not have bioactive compounds and are only used based on the resembling characteristic of the plant, rather than actual observed efficacy. It is unlikely that knowledgeable traditional healers, adept at discerning illness from health, would continue to utilize plants with patients that were not effective and did not provide healing benefits for their patients, merely because the plants physically resemble some part or characteristic of the human body.

Bradley Bennett (2007) proposed that the Doctrine of Signatures should not be considered so much an explanation for plant discovery, but a means for disseminating knowledge. The Doctrine of Signatures is a mnemonic device and therefore very valuable in traditional cultures around the world, as a means for sharing and remembering information regarding traditional plant remedies, and therefore preserving traditional knowledge. All four species with characteristics related to the Doctrine of Signatures from this study (two with domatia shaped like human body parts, and two with leaves with purple undersides), displayed novel estrogenic bioactivity in this study.

5.3 Q'eqchi' Women's Healing Traditions in Belize

My third hypothesis was: *women's healing traditions are being lost in the Q'eqchi' communities of Southern Belize due to a combination of factors related to their emigration from Guatemala and the socio-cultural setting of Belize.* Everyone involved with the study agreed that in the past there were more female healers than there are today. Of the three Q'eqchi' midwives I found in Toledo, only two were practicing. One had recently moved 5 years earlier to Belize from Guatemala, and wanted to teach midwifery to interested women in Belize but was afraid to. The other was afraid to meet with me and only had the interview with my Q'eqchi' research assistant.

Many months into my fieldwork and I was still told by everyone I had met that no female Q'eqchi' healers existed in Belize. Some researchers may have stopped there, and continued their work only with the male healers. It is important to consider the bias this would bring to the data, meant to be representative of Q'eqchi' traditional medical knowledge. In cultures where women are not recognized, or no longer recognized as healers, traditional medical knowledge of great worth for their community still exists that is specific to women. Undoubtedly, many studies are missing women's traditional knowledge, if it is not formally recognized at the community level.

Women's healing exists "informally" in other societies as well. Although none of these situations parallel that of women's healing traditions in the Q'eqchi' Maya communities of Belize, the examples collectively shed light on the many factors affecting such traditions. In Sri Lanka, local-level female healers reside in nearly every community, and are the first choice beyond home remedies for primary health care. However, unlike other various classes of healers in Sri Lanka, these women are not recognized as healers beyond the community, and they and their practice are not named. They therefore do not appear on surveys or research on health care providers even though they are recognized for their healing ability at the local level and widely sought after for their treatment. In Sri Lanka formalized, institutionalized healing is male-dominated, whereas noninstitutional healing at the local level is usually the domain of women who tend to work out of their homes (Nordstrom 1989).

Among the Saraguro people in the Ecuadorean Andes, women assume the role of family healers after their first successful pregnancy. Mothers gain experience through treating their family members and increase their knowledge through discussions with female friends and relatives, and some experimentation. This knowledge is passed specifically from mother to daughter. By the time a woman's daughters reach puberty they are their mother's assistants in healing the family. Traditional herbalists in the community do not discuss remedies with one another as mothers do, so mothers often have a more extensive knowledge and more eclectic understanding of health (Finerman 1989).

In the Chinantec communities of highland Oaxaca, Mexico, Carole Browner (1989) presents a situation in which the services of a woman trained in a government midwifery training program, were not utilized by women in the community in most cases. The midwife's knowledge of biomedical midwifery was recognized throughout the community, however her social status was not enhanced by her work, nor was she asked to attend most births. Browner found that most women in the Chinantec community felt they could manage uncomplicated births at home without the help of a midwife, especially if they had already given birth without the assistance of a midwife. Since there are many women in the community who do not consider themselves midwives but have equivalent knowledge to the midwives, the services sold by government trained midwives will not be in great demand. So there is little advantage to holding this role in the Chinantec community, and therefore does not provide much interest as a career path for women (Browner 1989).

The impact of the loss of traditional midwives in Belize is great when you consider that women in remote villages of Toledo are giving birth at home alone with untrained family members only, when traditionally they would have the assistance of a traditionally trained Q'eqchi' midwife. The depth of women's ethnomedical knowledge will be lost with the passing of the last generation of traditional Q'eqchi' midwives in Belize. Besides losing the medicinal plant knowledge specific to women's healing traditions, other consequences of the loss of women's traditional knowledge can be foreseen. In the Bolivian Amazon, among the Tsimane', a horticulturalist and foraging society, mothers with higher levels of local plant knowledge and use have healthier children (McDade et al. 2007). Not only do they have knowledge about remedies to treat their children's illnesses, they may also provide them with more diverse plant-based foods that better supply the macro and micronutrients needed for growth and development. Ethnobotanical knowledge is similarly important for combatting malnutrition and under-nutrition in underdeveloped areas such as the Toledo District of Belize, which has a significantly higher rate of growth retardation (39%), compared to 15.4% in Belize overall (Pan American Health Organization 2002).

Working toward the complementary coexistence and support between local traditional medical systems and biomedicine is critical for providing for the healthcare needs of the population of people in the developing world. Allowing traditional midwives to practice and providing MOH trainings that are culturally sensitive allowing for traditional practices, while training women in the necessary skills is needed. This will encourage women to maintain a working relationship with the MOH. Traditional medicine and biomedicine can coexist, and overlap or complement one another to suit the specific needs of the community. In a rural Quechua community in Bolivia, Vandebroek et al. (2008) found that biomedicine and traditional medicine overlap in the treatment of some health conditions including respiratory infections, fevers, wounds, and bruises, while traditional medicine appeared to be complementary to biomedical care for chronic illnesses, especially arthritis.

The use of health care facilities for birthing is low among indigenous women in the rural Peruvian Andes, partly due to cultural insensitivities in the health care system. A culturally sensitive delivery care model was implemented for Quechua women including adding a rope and bench for vertical delivery position, inclusion of family and T.B.A.s in the delivery process, and the use of the Quechua language. Since the implementation of this model, use of the health care facility has increased from 6 % in 1999 to 83 % in 2007, with high satisfaction among the mothers (Gabrysch et al. 2009).

Vandebroek et al. (2008) discuss how traditional medicine (TM) is both "traditional" and "modern." TM is traditional in the sense that it is in agreement with and inclusive of local belief systems and explanatory models of illness. At the same time, TM is "modern" or "adaptive" for being all-round and dealing with a diverse array of local health conditions." Quinlan and Quinlan (2007) discuss modernization and medicinal plant knowledge in a Caribbean village on the island of Dominica:

> We do not conclude that modernization leaves traditional medicine unchanged; rather, that the nature of the change may be subtle, complex, and specific to particular treatments. Perhaps, rather than assessing loss of reliance on phytotherapies, we should be looking for more detailed information. Ethnomedical treatments might be more likely to change than to disappear. New herbs, or new use of herbs, may enter communities, while less effective treatments that are harder to grow or species that are more difficult to harvest may fall from traditional use.

Plant use among the Belizean Q'eqchi people is certainly changing and adapting. It may be becoming less gendered, as women healers disappear, and interested young women are more likely to learn about medicinal plants from men such as their husbands or male family members such as fathers or uncles. There are still knowledgeable elder Q'eqchi' women however, and they can be encouraged to share their knowledge with other women. I facilitated the formation of a Q'eqchi' women's herbalism group, in which elder women are invited to give an informal workshop on medicinal plants they use to other women interested in learning more (Fig. 5.2).

Programs like the women's herbalism group are needed to facilitate the revitalization of healing traditions in the Q'eqchi' communities of Belize, and can be successful considering that Q'eqchi' youth still maintain the interest and desire to learn traditional plant-based remedies and ethnomedical treatments. Without these opportunities for learning, women's traditional medical knowledge will disappear when the eldest generations pass. Among the Q'eqchi', this study has shown that

Fig. 5.2 Maya women's herbalism group meeting in Punta Gorda Town

men have knowledge of medicinal plants from forested areas where habitat is rapidly declining. Q'eqchi' women's knowledge of healing plants may be more adaptable to such environmental changes, as the species they utilize can be found commonly growing in disturbed areas, are grown in home gardens, or used in the Maya cuisine.

Future studies focusing on other branches of traditional Q'eqchi' medicine, in addition to this study on reproductive ethnomedicine, with a comparative look at male and female knowledge, are needed to comprehensively document Q'eqchi' ethnomedical wisdom. These traditions were once maintained primarily orally, however with major factors including globalization, geographic displacement, and the ongoing marginalization of indigenous peoples, these rich traditions have become widely fragmented. The desire to document, preserve, and further validate traditional knowledge with the aim of educating the younger generations and revitalizing ethnomedical practices is increasingly coming from indigenous communities themselves. Research that is aligned with the objectives of the communities we work with is crucial for identifying areas of traditional knowledge that are being lost most rapidly. This is necessary for identifying the communities' own priorities for stabilizing the global loss of traditional knowledge at the local level, where the effects can be felt viscerally by those people whose health and livelihoods depend upon the sustainable utilization of the natural world.

References

Bennett B (2007) Doctrine of signatures: an explanation of medicinal plant discovery of dissemination of knowledge? Econ Bot 61(3):246–255

Browner C (1989) The management of reproduction in an egalitarian society. In: McClain CS (ed) Women as healers: cross-cultural perspectives. Rutgers University Press, New Brunswick

Centers for Disease Control and Prevention (CDC) (2013) http://www.cdc.gov/reproductivehealth/WomensRH/00-04-FS_Hysterectomy.htm. Accessed 15 Feb 2013

Finerman R (1989) The forgotten healers: women as family healers in an Andean Indian community. In: McClain CS (ed) Women as healers: cross-cultural perspectives. Rutgers University Press, New Brunswick

Fuchs-Young R, Howe S, Hale L, Miles R, Walker C (1996) Inhibition of estrogen-stimulated growth of uterine leiomyomas by selective estrogen receptor modulators. Mol Carcinog 17:151–159
Gabrysch S, Lema C, Bedriñana E, Bautista MA, Maica R, Campbell OMR, Miranda JJ (2009) Cultural adaptations of birthing services in rural Ayacucho, Peru. Bull World Health Organ 87:724–729
Lang JH, Zhu L, Sun ZJ, Chen J (2003) Estrogen levels and estrogen receptors in patients with stress urinary incontinence and pelvic organ prolapse. Int J Gynecol Obstetrics 80:35–39
Manderson L (1981) Roasting, smoking and dieting in response to birth: Malay confinement in cross-cultural perspective. Soc Sci Med 15B:509–520
McDade TW, Reyes-Garcia V, Blackinton P, Tanner S, Huanca T, Leonard WR (2007) Ethnobotanical knowledge is associated with indices of child health in the Bolivian Amazon. PNAS 104(15):6134–6139
Nordstrom C (1989) It's all in a name: local-level female healers in Sri Lanka. In: McClain CS (ed) Women as healers: cross-cultural perspectives. Rutgers University Press, New Brunswick
Pan American Health Organization (2002) "Belize." Health in the Americas
Quinlan MB, Quinlan RJ (2007) Modernization and medicinal plant knowledge in a Caribbean horticultural village. Med Anthropol Q 21(2):169–192
Rizner TL (2009) Estrogen metabolism and action in endometriosis. Mol Cell Endocrinol 307:8–18
Sich D (1981) Traditional concepts and on pregnancy, birth and post partum period in rural Korea. Soc Sci Med 15B:65–69
Vandebroek I, Thomas E, Sanca S, Van Damme P, Van Puyvelde L, De Kimpe N (2008) Comparison of health conditions treated with traditional and biomedical health care in a Quechua community in rural Bolivia. J Ethnobiol Ethnomed 4:1
Ward RL (1977) Curing on Ponape: a medical ethnography. Ph.D. Dissertation, Tulane University
Whiteman MK, Hillis SD, Jamieson DJ, Morrow B, Podgornik MN, Brett KM, Marchbanks PA (2008) Inpatient hysterectomy surveillance in the United States, 2000–2004. Am J Obstet Gynecol 198(1):34.e1–e7

Appendix

Appendix I. Interview Questions

Date?
Location of Interview?
Sociodemographic Data
Name?
Age?
Gender?
Ethnicity?
Religion?
Marital Status?
Children (Ages/Genders)?
Birthplace?
Current Residential Location (Village/District)?
Sociodemographic Details
Where were your parents born?
How many siblings did you have (#sisters/#brothers)?
What sort of schooling have you completed?
How long have you lived in Belize?
How long have you lived here?
What languages do you speak?
What is your primary language?
What is your occupation?
Medicine
What type of healer do you call yourself?
Who taught you what you know about healing?
Where did you learn?
Has anyone else in your family worked as a healer?
How long have you been working as a healer?
When did you know that you had the ability to help people in this way?
Is this a part-time or full-time occupation for you?

J. De Gezelle, *Q'eqchi' Maya Reproductive Ethnomedicine*,
SpringerBriefs in Plant Science, DOI 10.1007/978-3-319-10744-8

Do you do anything else in addition to make a living?
How many people seek you for treatment per day (or per week)?
Do you treat both women and men?
Do you treat individuals outside of your family?
How do people hear about your healing practice?
Do you use plants?
Do you also use store-bought medicines?
Do you use any other methods in your healing practice?
Plant Usage
Where do you get the plants you use?
Are there specific times of day, times of week, times of month, season, etc. that you collect plants during?
If so, could you explain why?
Are there specific times of day, times of week, times of month, season, etc. that you administer plants during?
If so, could you explain why?
Estrogen-Related Conditions and Regulations
(Dysmenorrhea) How do you treat a woman who has painful menstruation?
(Menorrhagia) How do you treat a woman who has very heavy menstruation?
(Primary amenorrhea) How do you treat a girl who has not begun to menstruate by her mid to late teenage years?
(Secondary amenorrhea) How do you treat a woman whose menstruation stops for a few months or more?
(Irregular menstruation) How do you treat a woman whose menstruation is very irregular?
(Infertility) How do you treat a woman who is unable to get pregnant?
(Menopausal hot flashes) How do you treat older women who no longer menstruate and complain of hot flashes?
Plant-Based Treatment Specifics (for each plant mentioned for the above listed conditions, the following questions will be asked)
What do you call this plant?
Where does this plant grow?
Where do you collect this plant?
Do you encourage this plant's growth in any way?
Are there specific times of day, times of week, times of month, season, etc. that you collect this plant for this condition?
Are there any special procedures or tools you need to collect this plant?
What part of the plant do you use?
Have you seen this plant with flowers or fruit?
If so, what time of year and where?
How do you prepare the plant to treat this condition?
How do you administer the treatment for this condition?
Where did you learn this treatment?
How effective is the treatment?
How long before you start to see the patient's condition improve while taking this medicine?

How do you tell if this treatment is working to heal the patient?
How does this medicine work to heal or treat the patient?
Can the plant be stored for later use or do you need to use it immediately after collection?
If you do store it, how is the plant stored and how long will it last?
How long have you used this plant as a medicine?
How far do you usually have to travel to collect it?
Have you always had to travel this far?
Has your use of the plant or its effectiveness changed over the time you have been using it?

MIX
Papier aus verantwortungsvollen Quellen
Paper from responsible sources
FSC® C105338

If you have any concerns about our products,
you can contact us on
ProductSafety@springernature.com

In case Publisher is established outside the EU,
the EU authorized representative is:
Springer Nature Customer Service Center GmbH
Europaplatz 3, 69115 Heidelberg, Germany

Printed by Libri Plureos GmbH
in Hamburg, Germany